D0149251

Facing Climate Change

Facing Climate Change

Columbia University Press   *New York*

An Integrated Path to the Future

Jeffrey T. Kiehl

Columbia University Press
Publishers Since 1893
New York   Chichester, West Sussex
cup.columbia.edu
Copyright © 2016 Columbia University Press
All rights reserved

Library of Congress Cataloging-in-Publication Data
Kiehl, J. T. (Jeffrey T.)
Facing climate change : an integrated path to the future /
Jeffrey T. Kiehl.
New York : Columbia University Press, [2016] | Includes
bibliographical references and index.
LCCN 2015022596 | ISBN 9780231177184 (cloth : alk. paper) |
ISBN 9780231541169 (e-book)
LCSH: Climatic changes—Psychological aspects. | Global
environmental change—Psychological aspects. | Environmental psychology. |
Human ecology—Psychological aspects.
LCC BF353.5.C55 K54 2016 | DDC 155. 9/15—dc23
LC record available at http://lccn.loc.gov/2015022596

⊗

Columbia University Press books are printed on permanent and durable
acid-free paper.
This book is printed on paper with recycled content.
Printed in the United States of America

c 10 9 8 7 6 5 4 3 2

COVER IMAGE: Photo by Jeffrey T. Kiehl
COVER DESIGN: Milenda Nan Ok Lee
All text photos by Jeffrey T. Kiehl

References to websites (URLs) were accurate at the time of writing. Neither the
author nor Columbia University Press is responsible for URLs that may have
expired or changed since the manuscript was prepared.

For Nancy, Kate, Alexis, and Matthew

We are now faced with the fact that tomorrow is today. We are confronted with the fierce urgency of *now*. In this unfolding conundrum of life and history there is such a thing as being too late. . . . We may cry out desperately for time to pause in her passage, but time is deaf to every plea and rushes on. Over the bleached bones and jumbled residue of numerous civilizations are written the pathetic words: "Too late . . ."

—Martin Luther King Jr.

—————————————

As any change must begin somewhere, it is the single individual who will experience it and carry it through. The change must indeed begin with an individual; it might be any one of us. Nobody can afford to look round and to wait for somebody else to do what he is loath to do himself. But since nobody seems to know what to do, it might be worthwhile for each of us to ask himself whether by any chance his or her unconscious may know something that will help us.

—C. G. Jung

Contents

Contents

Preface

Our demand for energy and our consumption of Earth's natural resources have pushed the planet into a state of great peril. Becoming aware of this environmental crisis is a first step toward transformation. It is natural to experience overwhelming distress when contemplating this situation. Beneath the fear of losing things close to us is the more fundamental fear of losing the world that has given birth to life itself. I sense this deep fear whenever I talk with people about global warming. The seeds of change are embedded in this generalized anxiety. Crises often give birth to creative transformation. Opportunities to create a better world can arise from our current situation, in which we are called to participate actively in this transformation. I have explored ways to begin the journey toward a flourishing future. The paths of science, Jungian psychology, philosophy, and Buddhism provide practical tools to understand the barriers to transformation and, more importantly, the means of breaking through them.

Science provides us with a clear picture of how and why Earth's climate is changing. The facts, based on observations, lead to the conclusion that we are altering the climate. The implications for the future are also certain. We cannot continue our current behavior toward the planet's precious natural resources. We must begin to develop a more caring relationship toward the world. Although technology will play an essential role in addressing our problems, we must confront the facts of the fundamental psychological roots of this problem. Facing *psyche* is a pivotal point for our transformation. By psyche, I mean both the conscious and unconscious processes that make up our psychology.

Jungian psychology helps us understand why we fear the changes looming in our future. In gazing into the depths of the unconscious, we see how it is possible to deny the existence of threats in spite of their reality. Jung discovered that our psyche contains coherent patterns of behavior that are charged with affect. Jung called these feeling-toned structures complexes. Complexes not only occur in individuals; they appear in entire cultures. Jungian psychology helps us analyze how individual and cultural complexes contribute to the existence of our environmental situation. Awareness of these coherent emotional patterns opens us to developing a more meaningful relationship with them, allowing us to move through our habitual patterns of negativity. Breaking through the psychological barriers to change brings us to a threshold of transformation. On a deeper psychological level, Jung proposed the idea of archetypes, which are universally shared patterns of perception. Archetypes appear across cultures in the form of images and metaphors. Perhaps the most universal archetypal images are those of mother and father. Archetypes are the shared lenses through which we view the world.

Philosophy prompts us to ask questions concerning our way of being in the world. Throughout this work I use the word "world" as a signifier not of planet Earth but as the particular environment we experience. Thus, we have an inner psychological world, an outer everyday world, a social world, and a global collective world. In asking questions such as why we are here or how our being affects the world, we become aware of how we are a part of the world. In our explorations of the phenomenal world, the felt-sense world we experience in our everyday lives, we can begin to develop an ability to see deeply. This is a way of sensing the beauty and richness of things around us. We not only see the surface of things but can appreciate the inherent value of what is in the world. Sensing the world this way places us in a closer relationship to all things. This type of focusing grounds us so that we can see the world through a lens of fearlessness rather than fear.

Buddhism is a path that grounds us in awareness. I choose this particular path because it feels right for me. There are, however, many roads to finding inner stillness. Many religions have practices that open one to the transpersonal, and in connecting with it we awaken to our true interconnectedness with the world. Here, the word "transpersonal" means an awareness that transcends the ego's perspective. This awareness recognizes that there is more to the world than our individual existence. What transcends the ego perspective is rooted in spiritual experience, be it God, the Tao, Buddha-nature, or nature itself. An experience of the transpersonal provides us with meaning, which seems so elusive in a world rooted in the view that we are fundamentally separate individuals. From a Buddhist perspective, our ability to experience a transpersonal dimension in life ultimately depends on how we relate to mind. By becoming more mindful, we are able to change our presence and purpose in this world, and our actions will thereby arise from a sense of compassion for all beings.

These are the threads that I believe are critical for weaving our new tapestry of creative transformation. I feel that by bringing head and heart together we will be able to create a sustainable, flourishing future. History shows that humanity can rise to meet great challenges and overcome seemingly insurmountable odds. There is no question that the world is mired in continual conflict, from the local to global scales. It is difficult to see how we can transcend this state of turmoil. However, our basic nature of empathic, compassionate being can unite us in the creation of a new world of cooperation, creativity, and connectedness. I invite you to join me in weaving this tapestry of a flourishing future.

Acknowledgments

My journey in writing this book began over ten years ago, in planning a workshop for the Aspen Global Change Institute called "Exploring Boundaries of Nature: A Reflective Dialogue on the Environment." My involvement in AGCI workshops continues to evoke reveries about our deep connections to Earth. I thank John Katzenberger, the director of AGCI, for his years of interest in my work. I thank Steven Bennett of Regis University for introducing me to phenomenology. I thank the Jungian analysts Bernice Hill, Jerry Wright, David Schoen, and Susan Olsen for heartfelt support. I thank James Hurrell and William Large of the National Center for Atmospheric Research for their support over the years. I thank the Advanced Studies Program of NCAR and Lisa Sloan for making my year visit to UC Santa Cruz possible. My visit to Santa Cruz allowed me to open up to the writing process. I thank Bill Herr at Esalen, who awakened me to the power of sharing one's writing with others. I am indebted to Andy Couturier of The Opening in Santa Cruz for allowing me to find my true voice. I thank fellow members of Andy's Book Completion Group, who provided so much encouragement. I thank John Laue, poet extraordinaire, for his guidance. I am deeply thankful for John Cunningham's editorial assistance and friendship. John's patience and tender care made the book what it is. I thank Peter Jones of the Trident Café for his constructive comments. I offer tremendous gratitude to the dharma teachers and the sangha members I have known over the past forty years and give a deep bow to the Boulder Zen Center, the Cambridge Buddhist Association, and the Boulder Shambhala Center. I thank Patrick Fitzgerald of Columbia University Press for his support and belief in my work. The staff at Columbia University Press, including Kathyrn Schell and Michael Haskell, were wonderful to work with. I also thank Robert Fellman for his excellent editing of the manuscript. I deeply appreciate the comments from Ben Santer, a scientist of tremendous integrity. Much of this book was written in the early mornings, sitting at the Trident Café in Boulder, Colorado, and at Verve Coffee in Santa Cruz, California. I thank the staff of these two soulful places for providing an inviting writing environment. This work was finished during a personal retreat at Esalen Institute in Big Sur; the quiet, sacred space of Esalen is a constant reminder of why we need to preserve the beauty of this world. My writing was supported in part by funds received from the AGU Climate Communication Prize. I thank my parents, Bob and Alma, for their faith in me. I am forever thankful for the unwavering support of my wife, Nancy, and my daughters, Katelyn and Alexis. I am thankful for Matthew, my son-in-law. Ultimately, this work is for Kate, Lexi, Matthew, and future generations; may we all walk the Blessing Way into a flourishing future.

I

Changes

A Journey from Climate Science to Psychology

The goal of life is to make your heartbeat match the beat of
the universe, to match your nature with Nature.

—Joseph Campbell

My wife and I walk up a trail in the foothills, a trail I first
traversed some forty years ago, when I arrived in Boulder. I
have walked this path many times since. The contour of the
trail is familiar, as is the vegetation bordering it . I have been
away from this particular trail for a few years, and as we walk
I am struck by the radical changes that have taken place. Large
boulders lie strewn across a flat streambed. The trees that once
grew on either side of the narrow stream have fallen on their
sides. Recently, Boulder endured extreme flooding, altering
its landscape in many places. The flood was caused by a record amount of rainfall over a few days. I am struck by the
formidable effect that the floodwaters have had on this once-
familiar terrain. The place I have known for so many years is
now changed.

Life is full of change, twists and turns, ascents and descents;
my life is no exception. I remember the day when my professional trajectory changed significantly. Sitting in my office, I
was looking at the carbon dioxide levels projected to occur by
the end of the twenty-first century, assuming we continue on

our "business-as-usual" path. These levels will be more than three and a half times what they were before the Industrial Revolution. The projected global warming from this increase in greenhouse gas will be approximately 6 degrees Fahrenheit. I thought about what these numbers would mean to the person on the street. What was the context for these carbon dioxide levels? When was the last time that there was this much carbon dioxide in the atmosphere, and what was the climate like then? It turns out you would need to go back roughly 35 million years to find these levels of carbon dioxide! Could it be possible that in a mere ninety years we will return Earth's atmosphere to a state not seen since the deep geologic past? I began to feel uneasy and realized that this was a path we could not pass on to future generations. This was the beginning of my turning.

I leaned back in my chair, realizing society had to change to avoid the worst consequences of our actions toward the planet. I asked myself: why do we choose to continue on this destructive path? After all, scientists had been pointing out the dangers of staying the business-as-usual course for decades. The basic science on the issue was over a century old, and since then increasing information has repeatedly confirmed that humans are altering Earth's atmosphere and climate. The science of global climate change is summarized in the box at the end of this chapter.

Despite the overwhelming scientific evidence of human effects on climate, little had been done to address the problem. There was a clear disconnection between knowledge and action. I wanted to understand the origins of this dilemma. *Why do we choose to do so little about this problem, when we know so much about it and when so much is at stake? Why do we resist the facts about this issue?*

Attempting to answer these questions seemed daunting. I knew many people were studying the barriers and impediments to action on climate change, but I felt that the role of psychology was receiving little attention. Since action on climate change required shifts in behavior, I was convinced that psychology had something to offer in answering questions about the causes of climate change denial and about the roots of resistance to changing our behavior. I knew I would need to change course to explore the psychological dimensions of the issue.

The thought of going back at age fifty to study psychology was a challenge, but I felt I had to do it. With the approval of my research institution, I went back to school and earned a master's degree in clinical psychology. I chose clinical work because I wanted to understand human behavior at the most practical level. During my training, I worked with individuals who struggled with life. Many lived self-destructive lives, unable to change. I began to see parallels between their lives and our reluctance to confront the climate crisis.

My educational journey was nontraditional: I chose to study Jungian psychology, an approach currently in large part dismissed by academia, as I felt Jungian psychology had much to offer in helping explain why we are in our current situation regarding global warming.

The basic tenets of Jungian psychology are that the unconscious plays a critical role in determining our behavior and that it contains processes that often conflict with our conscious choices. Jung also felt that the unconscious contains coherent, emotionally charged forms of perception called archetypes. These universal patterns of perception developed over evolutionary timescales. Interestingly, these assumptions are supported by current neuroscience. Jung believed that the unconscious worked at the level of image and metaphor, a belief that is also supported by cognitive research. Jung had a profound appreciation of how important the environment is to our well-being. He was interested in how we relate to the natural world and in what has separated us from it. As early as the 1930s he expressed concern about environmental degradation and overpopulation. I resonated with Jung's views of psychology and his deep appreciation for the natural world.

In addition, Jung had a great appreciation for the role of spirituality in the lives of most human beings on the planet. This is perhaps the main reason his ideas have been marginalized by mainstream academia, yet it is undeniable that spirituality plays an important role in life. Being in relationship with transpersonal experiences helps us live a better life, and often these experiences involve a deep connection to the natural world. Frankly, I find the dismissal of spirituality counterproductive to establishing meaningful dialogues about the problems facing the world. I feel it essential to include spirituality, morality, and ethics in the discussion about climate change.

As a Jungian analyst, I have assisted people with their personal search for meaning. I have also used a Jungian lens to study the issue of global warming. When I give presentations on this issue, both in this country and others, I find that people get excited by how this perspective brings fresh understanding to the problem. A Jungian psychological approach allows us to see problems in terms of affective images and metaphors. Our lives are filled with narratives thoroughly interwoven with image and metaphor, so climate change narratives rooted in image and metaphor touch both mind and heart.

Studying psychology also opened me to seeing new ways to communicate climate science to the public. Talking about climate change is like discussing evolution, genetically modified foods, or birth control. These all are highly emotional issues. Studying how people make decisions around emotional issues opened my eyes to why communicating the science of climate change is so challenging. I began to integrate my psychological knowledge into public presentations, and I spoke with climate scientists about how to work psychological concepts into their talks. Improving how scientists communicate to the public is essential, but that alone is not the complete solution to changing our current path. The psychological and sociological barriers around climate change are immense, which is why I feel a Jungian psychological approach to understanding these issues is essential.

In terms of my own climate research, I turned to looking into Earth's deep past. Given that we are returning the atmosphere to a state not seen for tens of millions of years, I wanted to know more about past eras of dramatic climate change. In particular, I was interested in the Earth's five major extinction events and what role climate may have played in these. The beauty of past climates is that they not only include the physical climate system but also involve biology, geology, and ecology. The study of the natural world through science has been described as lifting the veil on nature. By studying deep time, we unveil Earth's intricate connections between life and climate.

In the midst of my paleoclimate research I recalled my childhood interest in dinosaurs. Like most kids, I was fascinated with dinosaurs and collected books and figurines of these majestic beings, though these could not fully satisfy my passion. Luckily, I grew up in Pennsylvania, the land of coal mining. My parents would drive me to great debris deposits produced by active mines, where I scrambled happily over these heaps of coal-bound history in search of fossils, which were in great abundance. It's clear why the name "Pennsylvania" holds a place in the geologic timeline. Everywhere I looked, there were fossils: four-foot-long fern fronds, remnants of ancient sea life, and intricately variegated tree trunks. After a day of plundering the mountain's debris, I would load up the car with treasures from the depths of time, and we would head home. Our basement became a neighborhood natural history museum, the entire floor covered in fossils. As you gingerly stepped around that basement floor, you gazed upon millions of years of Earth's history. My love for science began with those early "digs."

The further back in time we look, the greater the sweep of knowledge we gain. Perhaps of greatest importance is that gazing into past worlds and gathering our discoveries provides meaning for our present-day lives. Jung states, "If we are to see things in their right perspective, we need to understand the past of man as well as his present." Looking into the past tells us that by the time of the Industrial Revolution, humans had taken a formidable leap in their ability to disturb the environment. We had turned our well-honed adaptive skills to finding more energy to fuel our increasing appetites. The discovery of fossil fuels was a revolutionary step in our ability to alter the planet. In a few brief centuries, fossil fuels formed tens to hundreds of millions of years ago were brought to the surface for their stored energy. The decayed, compressed plants and ocean organisms held the Sun's energy from the distant past in the form of hydrocarbons, and their extraction and burning released this ancient stored solar energy for our use. The burning also released carbon dioxide back into the atmosphere. Imagine the power of human ingenuity: to undo in a few centuries what it took nature hundreds of millions of years to do! Our ability to transform the planet is captured in the proposed new term for our geologic epoch, the Anthropocene. Our view into the past provides us with a rich picture of our place in the world and our current role in how the world is changing. The messages from the past are both illuminating (in terms of providing a context for present-day climate) and disturbing.

We need to recognize also that the problem of human influence on the planet is about more than our changing climate. It is about how our behaviors affect all aspects of life. The dinosaurs that interested me so much as a child had no choice regarding their fate. The unexpected impact of an asteroid led to their demise. We, however, have a clear choice to make because we are the primary cause of the current climate crisis. Our increasing use of fossil fuels—and the planetary warming arising from large-scale fossil fuel use—affects the economy, food systems, our physical and mental health, the places we live, transportation, and many nonhuman species. For more than thirty years, people like Fritjof Capra have been telling us that the world is highly interconnected. We cannot separate the worlds of energy, economy, food, health, dwelling place, transportation, and quality of life. We cannot use a measure like gross domestic product (GDP) to inform us of the true healthiness of person and planet. We must begin to recognize the true interdependency of these many facets of life and work to insure that all these individual facets flourish in the future. Most people care deeply about their children and the future of this planet. Many care about the survival of all lifeforms on the planet. The only way we can insure a viable future is to root our actions in the recognition of this interconnectedness.

Rather than allowing thoughts of interconnectedness to become too heady, I feel we need to plant our feet firmly on the ground concerning this issue. My studies in psychology introduced me to how we relate to our everyday world, the world that is present before us at each moment. An entire field, phenomenology—meaning *the study of things as they appear before us*—considers how we experience the everyday world around us. How often do we take the time to slow down and really see what is before us? How often do we slow down to touch and smell what is before us? How do we listen to the world around us, the sounds outside my window or the person speaking to me? If we are going to root ourselves in connectedness, we need to learn how to connect to ourselves and the world in which we live. This is practical work, not the domain of heady, speculative theorizing. The good news is that we can start at any time and place. As I stand in line at a grocery store I can take the time to notice the sights, sounds, smells, and tactile sensations I am experiencing in the moment. I become aware and engaged with the richness of the world surrounding me, and in that moment I experience connection.

We need to cultivate multiple ways of knowing our world, which is why the arts and humanities are essential to healing our split with the world. Sadly, these disciplines are dismissed as not *useful* to the world. Often they are viewed as luxuries or nonessential to life. I am not speaking only of intellectual knowing but a knowing with body, heart, and mind. And here is the key message of what I have to say about our dilemma with climate change: *We need to bring heart into the world.*

My turning was not only in intellectual awareness; it went beyond thought and included how I felt about the world in which I live. Scientists are supposed to suspend their value judgments because they may cloud our ability to see things as they "really" are. In areas of pure science this is possible. Studying the behavior of an elementary particle (is it a wave or a particle?) is perhaps less likely to evoke a strong feeling of connection. Studying the fate of Earth is another matter altogether.

As I look back on life, I realize my journey has been one of including more ways of knowing. I began my career in the physical sciences. I worked with chemists to consider the interactions of climate and chemistry. I became interested in the human dimensions of climate change, which prompted me to study psychology. In my studies to become a psychologist I was introduced to the subject of how we perceive and experience the phenomenal world. Now, I find myself getting more in touch with the world I care about. I feel more for this world and its preservation.

––––––––––

Our transformation into thinking beings with the ability to change our environment is impressive. However, our journey, taking place over millennia, must now be transformed again. Our new approach to being in this world must weave knowledge and wisdom together into a more encompassing tapestry. When we see what is happening to the world, we feel sadness and fear. Our first impulse is to flee these disturbing emotions, but it is important to stay with these intense feelings. It is through an open heart that wisdom enters the world. We are not only relying on our heads to make decisions about the future; we are tapping into our heartfelt wisdom.

Hiking that dramatically altered foothills trail brought sadness to my heart. Things as I remembered them had been radically changed by the flood. I was keenly aware that in our warming world we could expect an increasing number of incidents like that. The lessons from the deep past supported this understanding. Somehow we lost touch with the earth and now tread heavily on it. How do we get in touch with the earth again? Will we continue to walk our old path, which will inevitably alter the planet? What keeps us from changing paths? How do we begin to see the beauty and value of Earth in all of its magnificence? Will we choose to travel the new path, opening our minds and hearts to great transformation? The meditations that follow are my attempts at answering these questions.

UNDERSTANDING HUMAN-CAUSED CLIMATE CHANGE IN FIVE STEPS

Whenever I give presentations on climate change I like to discuss the five fundamental facts that prove humans are causing the planet to warm. These facts are based on direct observations in the atmosphere, laboratory measurements, and a basic law of physics. These facts have been confirmed by the majority of scientists working independently, over many years, on understanding Earth's climate system.

1. *The amount of atmospheric carbon dioxide is increasing.* Direct observations of the amount of carbon dioxide in the atmosphere over the past fifty years clearly show this increase. Trapped bubbles in ice cores that contain carbon dioxide tell us that the carbon dioxide concentration began to increase at the time of the Industrial Revolution, and this ice-core record merges nicely with the more recent direct observations. The current concentration of carbon dioxide is now 43 percent higher than what it was just before the Industrial Revolution.

2. *This observed increase in atmospheric carbon dioxide is caused by the burning of fossil fuels.* We have two independent observations that support this conclusion. First, carbon and oxygen atoms come in "flavors." The flavors are determined by the relative number of

protons and neutrons within the nucleus of the carbon atom. The technical term for flavors is "isotopes." Fossil fuels are composed of lighter carbon atoms as compared to the heavier flavor of carbon atoms found in emissions from volcanoes. Observations tell us that the increase in atmospheric carbon dioxide is coming from the lighter atoms found in fossil fuels. Second, burning any material requires oxygen. You can't start a fire in a vacuum. Scientists have been measuring the amount of oxygen in the atmosphere, and it is declining at a rate that matches the burning of fossil fuels. Don't panic; the decrease in oxygen is very small, but it is measurable. Carbon dioxide emitted from volcanoes does not arise from combustion and thus cannot explain the measured reduction in oxygen.

3. *Carbon dioxide is an efficient greenhouse gas.* The Irish scientist John Tyndall showed in the 1860s that carbon dioxide is very effective at absorbing infrared energy, which is the type of energy emitted by Earth's warm surface. By increasing atmospheric carbon dioxide, we trap more of this energy in the climate system. Detailed laboratory observations of the efficiency of carbon dioxide's greenhouse effect have continued since the early work of John Tyndall and show that the absorption by this greenhouse gas does not level off with increased carbon dioxide concentrations. Observations from Earth's deep past also support this conclusion by considering the levels of atmospheric carbon dioxide and the associated warmth at those times.

4. *Greenhouse trapping of energy in the climate system warms the planet.* This results from the fundamental law of conservation of energy. If energy is not allowed to escape to space, then it remains trapped in the climate system. John Tyndall stated in 1859 that the greenhouse effect caused by carbon dioxide must warm the planet. Much of this trapped energy is held in the oceans. When we study greenhouse warming of the planet we must look at all the energy trapped and stored in the ocean. Observations of the *total* energy stored in the oceans show a continual increase over the past fifty years. There has been no slowdown or hiatus in the warming of the planet as a whole!

5. *The warming from a doubling or tripling of carbon dioxide is substantial.* Looking at past climates when carbon dioxide levels were much higher than today tells us how sensitive the climate system is to increases in atmospheric carbon dioxide. The 43 percent increase in carbon dioxide since preindustrial times has already led to a warming of ~1.6°F. If we continue to burn fossil fuels at their current rate, then by 2100, we will more than triple the amount of carbon dioxide above what existed before humans began burning fossil fuels. The last time that much carbon dioxide existed in the atmosphere was around 35 to 40 million years ago, a time of great warmth, when there were no polar ice sheets and the sea level was much higher. This was also a time when our Sun was less bright than it is today.

Theory and models are in agreement with these fundamental facts and can be used to explore the myriad connections within the climate system. When models include the emissions of carbon dioxide into the atmosphere over the twentieth and early twenty-first centuries, the predicted increase in the concentration of atmospheric carbon dioxide is similar to what is observed. This increase in simulated greenhouse gas leads to a trapping of energy that warms the planet, again similar to what is observed. The models also predict that much of the trapped energy is stored in the oceans, again in agreement with observations. Models are representations of the complex climate system and can be used to carry out "what-if" experiments. They serve as laboratories for climate research. What if we did not include the human input of carbon dioxide into the climate system from the burning of fossil fuels? We study this question by running our climate models with no human emissions of carbon dioxide. The result from these model simulations is that no significant warming occurs over the twentieth and early twenty-first centuries. Thus the only way to explain the observed warming is by including human emissions from the burning of fossil fuels. The model simulations substantiate the five facts of climate change. Note that these models are able to simulate the features of natural variations in the climate system.

If the warming were caused by natural variations, it should appear in the simulations without the burning of fossil fuels. Yet it does not. The five facts of climate change, in conjunction with the supporting model results, provide definitive proof that humans are the main cause of our warming planet.

Learning to Embrace Change

If you don't change direction, you may end up
where you are heading.

—Lao Tzu

I see a flock of pelicans flying in formation, gliding with grace and ease between the curved coast of sand and sea. These regal, gray birds fly with certitude, as if they have in mind a planned path. With a slight tilt of outstretched wing, they drift further out to sea, skimming a few feet above the undulating ocean surface. At times their bodies touch the cresting waves in search of sustenance below. The fluid motion of birds and sea manifests both certainty and change.

The Greek philosopher Heraclitus said, "The only thing that is constant is change." Indeed, life is change. Yet amid unexpected change some things seem to remain predictable, like the apparent transit of the sun across the sky each day. Repeating phenomena provide us with a secure sense of the known. We also experience surprises in life: unpredictable changes that make us feel less secure. Yet at the same time, the unexpected opens us to discovering the new and exciting. Uncovering the hidden things in the world opens us to a sense of wonder.

To deal with life, we need ready access to each of these modes of being—stability and surprise—without denying one at the expense of the other. True wisdom is knowing which mode is best in a given situation. The predictable provides a sense of stability. Openness to unexpected change provides flexibility, helping us feel more resilient in a changing world. The pelican flying along the coast instinctually balances between these two modes of being, poised to respond to unexpected fluctuations in the wind to stay on her charted course. We too must be comfortable with the known but ready to respond to the unexpected. Our resilience, and the world's, depends on our ability to balance creatively the known and unknown ways of being in the world.

———————

Why do we choose to continue and repeat certain behaviors, despite their calamitous consequences? Why do we turn away from any message warning us of these destructive habits? It is like being on the Titanic and knowing about the looming icebergs but choosing to stay a steady course. What keeps us frozen into our habitual ways of living? One answer to these questions is related to our essential sense of being, which is defined by known patterns of behavior. Once, during a therapy session, my patient, with fear in her voice, said to me, "But if I changed, who would I be?" Our way of being is defined by patterns of beliefs and behaviors and may be so ingrained in us that the thought of altering them fills us with anxiety, even though those patterns are destructive.

"Who would I be if I changed?" Many people are willing to live with the deleterious side effects of predictable patterns of behavior rather than experience and work through their fear of change. I believe this is where we are with the issue of climate change. We have collectively defined our sense of being as consumers, believing that to buy and own is integral to our social fabric. The side effect of this way of being is the need for enormous amounts of energy to fuel such a lifestyle. What would it be like to explore alternative ways of living and generating energy for a reinvented personal and global lifestyle? How do you feel when I suggest going down this new path of being in the world? Do you feel a tinge—or more than a tinge—of anxiousness? Recently, I spoke to a small group on the issue of climate change. Within just a few minutes of this meeting, a number of people became strongly resistant to what I was saying. Strong, negative emotions began to spread throughout the room. I realized how much fear was present. If the science were accepted, change would be required. The fear of having to change had overwhelmed the people sitting in the room. Given our tendency to fear change, this reaction is normal. An important first step to any change is to sit with questions like these and become aware of how you feel. Meditating on questions and the feelings that arise around them is an important part of initiating creative change.

Another way to look at our inability to change, in spite of *our wanting* to change, is that it indicates the presence of an opposing "other" within. I want to change my life, but some part of me wants to continue my old way of living. What is this other part of our self that constantly keeps us fixed in the old counterproductive ways? Since consciously we are unaware of its presence, it must reside in the unconscious. Could it be that this other part that wants us to stay fixed in our old behaviors is actually trying to protect us? Here the fear is not so much about not consciously knowing who we would become if we changed but more the yearning for certainty. The inner part of us seeks the repeatable and remembered. This intense need for a secure, stable base is rooted in our early development. As infants and young children we are very vulnerable and need a fixed base to count on, a place, physical and psychological, that is predictably present when we return to it. If such security is established early on, then later in life we become more resilient individuals. We possess an inner sense of security and self-reliance. If we do not experience such security early in life, then we are often anxious about change, and security becomes paramount in life.

Clinical psychologists have noted that when we are able to accept who we really are, then in that moment we are able to change. Denying ourselves, or being completely unaware of parts of ourselves, means we cannot accept our very being in the world. This lack of acceptance creates a sense of alienation from our self and from others. By reflecting on who we are in terms of behaviors, beliefs, and internal struggles, we begin to accept our whole being. In this process of accepting who we are, we need to understand more than what is on the surface. With regard to our feelings about change, we need to reflect on our beliefs and behaviors that may be keeping us rigidly stuck in old patterns.

Metaphorically, a tension exists within us between what Jung referred to as the old person, or *senex*, who wants to keep things rigid and fixed, and the inner, eternally young person within, or *puer*, who wants constantly to experience new things in life. The *senex* is most comfortable with a set predictable routine and yearns for absolute certainty in things. In making decisions, the *senex* hesitates until collecting as much information as possible yet still may be reluctant to decide. The positive side of the *senex* is that they carefully assess and weigh situations. They are likely to reflect on matters before making quick decisions. Wisdom is a quality often associated with the *senex*. The negative side to the *senex* is a rigidity that stifles and prevents us from adapting to important changes. The result of excessive hesitancy or avoidance to change prevents positive transformation for an individual or an entire society. Those who vehemently fought against civil rights legislation or regulations on air pollution are emblems of the negative side of the *senex*. The image of the negative *senex* is seen in the figure of the crotchety old man or doddering old fool.

On the opposite end of the spectrum is the *puer*, whose tendency is to rush into decisions with little or no information. The *puer* loves to be surrounded by the new and exciting. You will never find a *puer* in an antique store—unless "vintage" items have become the hot new thing! They thrive on exploring many things simultaneously and do not like to be stuck on any single one. Their perception of time is the now, not yesterday or tomorrow. Commitment is a concept that fills the *puer* with discomfort, for to commit to a single relationship or cause would pin down their quick-moving spirit too much. Interestingly, the *puer* is often attracted to the opposite because something within them yearns for stability, but once they get too close to the other who is seeking commitment they flee the relationship. This opposition to *senex* qualities is reflected in the desire to act out impulsively, rebel against authority, and break rules.

The eternal battle between these two archetypal patterns can cause us to waver forever in a state of complete inaction. Those who are resistant to change fear the unknown more than those who are open to change. Working creatively with the *senex* and the *puer* attitudes requires that we recognize both as having something positive to offer. We are also called to reflect honestly on the darker elements of these two qualities: if we do not consider these dark aspects we place ourselves at the mercy of their hidden designs.

Social psychology studies have found that people tend to fall into two general groups: those who place great value on personal security and those who value harmony with others. Security seekers also place a higher priority on authority and a hierarchical social structure. They see hierarchical authority as creating a more protective secure environment; harmony seekers value equality and shared responsibility. Decisions by harmony seekers are based on what is best for the group rather than what is good for any particular individual. How do these two seeker paradigms relate to our perception of change? The security seeker will most likely be suspicious of change and perhaps even fear it. The harmony seeker will consider whether the perceived change is beneficial to society; if change benefits the group, then it will be embraced. This duality of being is playing out right now in terms of how people view the threat of climate change. Fear will turn security seekers away from taking any action on climate change. If the security seekers experience fear, their instinctual reaction is to protect themselves against whomever they see as a threat. They resist change even in the face of factual evidence arguing for change. Ironically, we would expect those who resist change to be the quintessential conservationists, but in reality it is just the opposite. This cognitive dissonance arises from the power of value systems; for the security seeker, *self*-protection is more important than protecting a more remote environment. Harmony seekers are the opposite in their relation to the world. Their sense of inclusivity means they feel connected to the world and are highly receptive to knowledge about the world in general.

Ultimately, we need to establish a dynamic collaborative relationship between these two parts of our selves—the seeker of security and the seeker of the unknown—and give each their due. From an evolutionary perspective we can see the value of both approaches to the world. A community composed of these two attitudes will be more resilient than a community composed of only one, be it the seeker of security or the explorer of the unknown. Essentially, as individuals and as a society, we need to mediate consciously between these two ways of seeing the world and accept that both are integral to survival. The moment we consciously accept these parts and integrate them into our being is the moment we can creatively work with change.

These two archetypal patterns also relate to the ethical manner in which we view the world. We can view the natural world as present to supply resources for our own personal wants and needs, in which case we view the world in terms of its utilitarian value. The view opposite to the utilitarian is that nature has intrinsic value independent of how we view it; whether humans are present or not the world has inherent value. In terms of our two social patterns of behavior, the security seekers will view the world through a lens of utilitarian worth in service of their needs. In addition, often this group views the natural world as a threat to its well-being. As noted, research indicates that those who resist change are more sensitive to perceived threats from the world. The harmony seeker will be more inclined to view the world as having the right to exist for its own sake, which fits with research that indicates that these types of seekers are more inclusive.

Not surprisingly, these two value systems relate directly to understanding the divisiveness around the issue of climate change. If we feel that the material world's value arises only from us, then we have the power to decide what out there is of sufficient worth to be protected. However, if our view is that matter is intrinsically valuable, then we are not in a position to assign worth to the material world, for we have a deep feeling connection to this world. Given this felt connection to the world, we will be far less likely to abuse it.

Our fundamental relation to nature is rooted in our psychological view of the world. The difficulty in addressing the issue of global warming exists because of the competing ways we value the world as either there to serve us or as there to exist for its own sake. The extraction and burning of fossil fuels and the consequent warming of the planet exists because we hold a mainly utilitarian view of the world. This narrow view of the relationship between psyche and matter has profound implications for environmental problems like global warming.

These two paradigms may provide insight into the differences between how European nations view climate change compared to the United States. In talking with colleagues from Europe, an interesting difference exists around the value of the individual and the group. The United States was founded on the principle of individual freedom. We broke from England over issues concerning representation and independence. The ethos of the United States still holds to the rights of individuals over that of the group. Decisions concerning how best to respond to challenging social issues like climate change require the inclusion of whole societies and not just particular individuals. Nations whose ethos is rooted in individual rights above those of society will be reluctant to take governmental action on issues like climate change. My European colleagues describe how decisions are less likely to be made to benefit the individual or a specific group. The decision process attempts to bring benefit to the most, not the least.

Humans have reached an era in history, the Anthropocene, where they have become a major force in shaping the natural world. Humans also possess a level of consciousness that brings with it ethical and moral responsibilities. Our decisions as a society can extend far beyond our own group, which is especially true in our current age of globalization. Decisions about energy and consumption extend across the planet, connecting the developed and developing worlds. Our national decisions no longer just affect us but the entire world. Climate change is a good example of this. We want to live a certain way, one full of material comforts, and this requires energy and natural resources. We consume fossil fuels to provide this lifestyle. The result of these decisions is to increase the level of carbon dioxide in the atmosphere, which increases the greenhouse effect and warms the planet. This warming has now reached a level where we are seeing its effects: a greater number of intense storms, rising sea levels, and degradation of the biosphere. Ironically, the very lifestyle we value so much is now threatened by climate change. Continuing to burn fossil fuels over the next few decades will change our climate and the way we live in the world tremendously. If we continue to live as we have done, even greater changes will occur in the future, many of which are presently invisible to us. I believe it is important to learn how to relate to change constructively. In doing so, we can begin to develop behaviors and beliefs that lead to creation as opposed to destruction.

Not all change need be disruptive or destructive. The gliding pelican uses changes in air currents to visit new locations that may be rich in food. Its ability to adapt and explore the changing currents brings abundance to its life. Similarly, changes in our personal lives and social structures often produce extremely positive outcomes. On the personal level, changing our beliefs and behaviors can result in a healthier lifestyle, more loving relationships, and a more peaceful state of mind. Although we may be reluctant to face change, when we do our lives become richer. On the social level, being open to change often ensures the security and health of a whole social system, producing a better world for our all. *Transformation begins with the willingness to face the fear of change.* The global challenges that face us today, such as poverty, energy, climate change, and social justice, can only be addressed through positive creative change. If we choose to live according to stories of limitation, separation, and hopelessness, then we will reap the grim seeds of these beliefs. It is time to put these deleterious beliefs to rest. If we are to move toward a world of flourishing sustainability, we will need to create a new story. One of these paths to create the new story is to understand and become familiar with our feelings around change. We have a natural tendency to resist change, and there are good reasons for this resistance. However, we also need to learn to engage with the reality of change. Exclusive resistance to change does not prevent the inevitable.

How to begin this process? First, we can become aware of what is going on within us when presented with the possibility of change. Change can approach slowly or rush forward, as in the case of an impending illness. Change places us face to face with choices, and having to choose can be very uncomfortable. Choice initiates internal and external dialogues that may take time to resolve. No matter how quickly we encounter change, we can choose to slow down and consider how we feel about the situation. We can talk with others about how they perceive change. Second, in engaging with change we can become aware of any conflicts the change generates within us. Do we sense the presence of an inner "other" wanting to hold on to the security of the known? Is there a part of us that feels excited about leaping into the unknown? Is there anxiety around becoming part of the change? Asking these questions with openness and curiosity is a positive step in dealing with any change. We often resist asking these questions because we don't know the answers, which brings us to the third step in engaging with change: realizing that we don't need to know the answers to our questions before engaging with the process of change. Often a fear of being wrong prevents us from approaching complex questions. Our culture is so oriented toward success that we fear not having the *right* answers to questions. But in many instances there are no *right* answers, only possibilities. By asking questions about how we feel in the moment of change we begin a journey of transformation. At such an early point on this journey we don't need all the answers, only the openness to possibilities.

I have explored many social and psychological dimensions of how we perceive and react to change. We are changing the world, and many of these changes are not and will not be beneficial to humans and the planet as a whole. *We must transform ourselves to transform the world.* I have argued that to do this we need to accept all of our attitudes and inner patterns and creatively work with them so we approach life from a more holistic attitude.

Change is inevitable. We can either face it with fear or with fearlessness. If we choose to engage with the process of change we have an opportunity to create solutions benefiting the whole world. Adopting a fearless attitude toward looming change is but one thread of the new story we must weave together. We must understand the magnitude of the problems we face and the role we play in creating these problems. An understanding of these aspects of our situation is necessary for finding solutions. The pelican developed and learned to maneuver those air currents from an early age. We too can learn from our past experiences. We can learn about what we fear and move beyond it.

Facing Our Fears Associated
with Climate Change

Anxiety has an unmistakable relation to expectation.

—Sigmund Freud

I am giving a presentation to a church group. I look at the audience, especially the expectant faces of the younger people. I
have been asked to talk about climate change, to provide them
with the current scientific understanding of what is happening
in the world. I describe how the last two decades have been the
warmest on record, with an accompanying record-setting loss
of Arctic sea ice. I present the latest facts: that in the United
States the number of days with record high temperatures is
now more than twice that of record lows, that it is estimated
that one-third of land animals and over half of plant species
could lose their habitats over the next eighty years. As I look
out into the audience, I sense that something has shifted in the
room. The expressions on people's faces tell me that various
moods now populate our gathering. I have seen this before and
realize it is time to move on from describing the facts of global
warming. I pause for a moment to ask how people are feeling. There is absolute silence, as if they have never been asked
this question before. Then I see a hand rise slowly: a young
woman expresses how helpless she feels in the face of such

immensely challenging information. After her brave, heartfelt admission, others begin to raise their hands and share feelings about our changing world and what they fear losing. The moods expressed include sadness, hopelessness, anger, denial, guilt, numbness, and fear. We sit together in silence, holding the multitude of moods. Giving voice to these silent spirits inhabiting our hearts brings a certain warmth to the room. In sharing our feelings about these issues a door opens, connecting us. Our humanness, our ability to suffer loss in our world, is perhaps the very thing that will lead to our transformation. Our shared feelings evoke within us a profound depth of caring.

Over years of presenting the scientific facts on climate change I have come to expect tremendous emotions around this issue. This awareness caused me to shift from presenting only the scientific facts to allowing time for people to express their feelings about those facts. The emotional response of anger about what is happening to the world, the feelings of hopelessness, helplessness, and denial about our role in this, including feelings of dissociation, are all indicators of trauma. No wonder a presentation on the science of climate change is so difficult to hear—it is a traumatizing experience!

Today we face many such traumatizing issues. Fear-filled feelings inhabit much more than the issue of climate change. People worry about the state of the economy and have concerns around health and aging or about how safe they feel in the world. We express deep-seated fears and anxieties about being in such a complex world. What comes to the fore when I listen to people's emotional responses is the tremendous sense of felt and perceived loss in their personal worlds.

Whenever I consider our changing world I experience feelings of loss as well. I reflect on the rapid changes in a climate system beneficial to all life on Earth. I consider the loss of clean oceans and the threats to the existence of so many species. I also recognize that these losses extend far beyond my personal sphere into the wider world. It is important to stay with and reflect on these feelings of loss, for we often experience that initial inclination to drift away. Dwelling with loss places us in the realms of sadness, emptiness, and ennui. If we are not careful, these feelings may lead us into a state of listless torpor, or apathy, perhaps even dark melancholia. Even reading this may be making you feel listless, but the cure for this is to stay with the feelings encircling our sense of loss rather than following our natural inclination to flee, that feeling of "Get me out of here as fast as possible!" Rather than fleeing, I invite you to stay a while with our mutual sense of loss. If we flee from the feelings, we may lose an opportunity to find meaning embedded within them. Dropping down into feeling and experiencing the sense of loss ultimately brings understanding and resilience to our lives.

Loss opens us to strong emotions, bringing with it a sense of separation and loneliness. Fundamentally, we are caring and relating beings who need to feel connection, and loss creates unfinished bridges. In the moment of deep loss, we are thrust into a world of solitude and sorrow. Feelings of anxiety, loneliness, helplessness, and sorrow are unwelcome in life. They lead us away from happiness and a sense of comfort. Is it any wonder that we avoid dwelling on anything that allows these uninvited shades to enter our lives? Indeed, we have become adept at closing the door on them so that we don't feel so lonely in the world.

Loss also creates a yearning for what is lost. We hold an inner image of the lost object and continue to look for it in the outer world. We may even project the inner image of the lost one out onto the world, but this projection can never alleviate our yearning. We continue to search but inevitably fail.

In loss, the visible becomes invisible! An object of importance disappears. We fear losing things, especially things we value. Often our fear of loss creates a barrier to thoughtful action to prevent it. Loss ranges from the trivial to the traumatic. Think about the last time you lost something of small value to you and how this initial experience grew in importance. I walk out of my house and discover that I don't have my keys to lock the front door. I don't have my car keys either, so I am going nowhere. A sense of *dis-ease* settles over me. I feel panic rising within. Damn it, I need to get to that meeting, but I can't lock my house, and I can't drive my car. I run back inside the house, trying to remember where I had last seen my keys. A hectic, rattling mystery begins to unfold in this most inopportune moment. Soon, I am in a full-blown panic, and I lose it. My reaction to the missing keys is completely out of proportion to the situation. Now my panic prevents me from finding my keys! Of course, feelings of panic and fear are far more intense when we are threatened with the loss of something of great value: the people we love, a job that provides us with our livelihood, or our strongly held beliefs. Such losses can seem unbearable, and we may emotionally "lose it" for a very long time.

Psychologically, the moment we lose it, we become blocked and are in the grip of a complex. Complexes are emotionally charged, coherent patterns in the psyche that behave like autonomous actors within us. Their particular form is influenced by our experiences with parents and peers and the social norms that pervade our life. Complexes can be thought of as scripts that have been written specifically for us and that determine what we see and guide how we behave. Under given situations, these old scripts can be triggered, and we act out the role, which often can be very different from who we consciously think we are. With the right stimulus a complex bursts into our lives like an uninvited guest at a party. I begin to worry about my financial situation, and this starts playing the script composed by my money complex. This particular script or autonomous role taps into my fearful feelings about being unable to support myself and family. What began as a simple passing thought rapidly grows into rampant worry. In that moment, the complex or script has taken over my life.

Beyond the personal, social and conditioned factors governing the development of scripts, complexes also arise from instinctive or archetypal patterns within the unconscious. Such complexes are most apparent when large groups or whole nations become emotionally overwhelmed and possessed. Highly emotional reactions to news stories are often examples of such archetypally rooted complexes. Even on the personal level, our scripts can tap into these more deep-seated patterns. Behind our personal father complex—be it positive or negative—lies the larger image of the Father in our psyche. The important thing to realize about complexes is that they can grab us and take center stage. Have you ever become so upset with someone only to ask yourself later: what came over me? The answer is that a part of you was triggered into following an old script, a complex. When I became emotionally distraught over my lost keys, so emotionally distraught that I overreacted to the situation, I was in the grip of a complex. This actually prevented me from solving the problem of the missing keys. We never get rid of complexes, but we can learn to live with them.

I have described complexes in detail, for they are the bridge between the stimulus of incoming information and activity from the outside world and our emotional responses to that information and activity. Complexes, in a sense, determine how strongly we react in a given situation. Will I be mildly annoyed or emotionally overwhelmed when confronted with what is happening out in the world? When we feel disempowered upon hearing the facts about climate change, we are experiencing the power of a complex. Whenever I give presentations on climate change, people always express their powerlessness around the issue. They just do not know what to do. Our reactions to this information can be so strong that we are unable to act. More disturbingly, others can manipulate these inner scripts in order to evoke strong reactions in us. Advertising is based on the effective use of evoking emotional responses around particular products. We all have a money complex; we all have some concern about our financial situation. If we are told that taking on the problem of climate change will cause personal financial distress, we will fear the loss of job and income. This fear may be so great that we will be unable to contemplate rationally doing something about the issue. Since loss evokes strong emotions, complexes are naturally connected to it. There are times when loss evokes valid strong emotions within us, but what of situations where complexes lead to over-the-top responses? It is for this reason that we need to explore our feelings around loss and how complexes affect our response to loss.

Loss may be tragic, like the sudden death of a loved one or an unwanted separation or divorce. Loss can also occur in the form of betrayal. Loss can be experienced as a disjuncture between what we expect and what is. We feel the missing "other" in a very palpable and profound way. Our soul yearns, our hearts break, and our bodies tremble. An experience like this places us in the realm of darkness, where the shades of the missing dwell. Once the loss occurs we turn our eyes to the past; we look back in remembrance or with yearning. These are the most difficult losses to deal with because there is no going back, no way to change the past to recover what has been lost. In these situations we are thrown onto the path of mourning.

Unlike *experienced* loss, which pulls us into the past, *anticipatory* loss occurs when we expect a loss to occur in the future. Have you ever been in a relationship in which you feared your loved one would leave you? As a parent, have you looked into the future when your children would leave home? The emotions associated with these future events are in reaction to an anticipated separation or loss. The emotions experienced in anticipated loss can be as intense as those of experienced loss. They may imprison us in a state of inactivity, but the key difference is that with anticipated loss, we still can do something about it, for it has not yet occurred. The fact that we often do not act to change the future is evidence of the internal power of complexes associated with the fear of future loss.

The concept of anticipated loss can help us understand the feelings that arise around the issue of global warming and why some of us are unable to accept this reality. Accepting the reality of climate change would ultimately mean having to do something about the problem. For those heavily invested in fossil fuels, any action to reduce the use of fossil fuels opens them to anticipated financial loss. The anxiety associated with this loss triggers defenses to reduce these strong feelings. Rationally, we know that fossil fuels are limited resources and that the cost of extracting them is going to increase. We also understand the environmental consequences of the continued use of fossil fuels, yet this understanding is negated by the anxiety associated with the perceived loss of the bountiful income associated with these fuels. A reasonable business strategy would be to develop new forms of energy that ensure continued financial stability with accompanying environmental benefits. This approach, however, is overridden by the deeper fears of loss. Illogical? Yes—but understandable from a psychological perspective.

Another perceived loss related to inaction on climate change involves those who are invested in notions of individualistic autonomy, who fear the loss of their basic right to choose how they live. Truly to address the issue of climate change will require some form of government intervention; the problem is so large that it will need national and international coordination. Legal decisions leading to personal limitations on how and what we consume are tremendously threatening to those who, above all else, value autonomy.

Some whose religious beliefs view God as the one divine being solely capable of determining Earth's climate will fear a loss in faith were they to accept the fact that humans are changing the climate. Perhaps the most pervasive anticipated loss regarding action on climate change is the fear of losing such basic needs as job security or the ability to provide for one's family.

What emotions are associated with these forms of loss? *Experienced* loss precipitates grief, sadness, and mourning. *Anticipated* loss generates anxiety, fear, and many of the emotions that arose in, for example, my meeting with the church group. All of these emotions threaten to overwhelm us. In experienced loss, we go through a natural grieving process, and in many cases we heal. In anticipated loss, we find ourselves in a liminal place of anxious anticipation in which our imaginations run wild. These emotions can incapacitate us and make us overly vulnerable, and with such persistent anxiety we are unable to function in the world. Luckily, however, means exist to avoid such emotional impairment. In moments of intense emotional instability, defense mechanisms arise within to restore a semblance of stability.

I'd like to pause and gather together these reflections on loss. When presented with the possibility of loss, we have a natural tendency to react emotionally. We become anxious about the future. These emotions are guided and amplified by our past conditioning and cultural values, which have created scripts, or complexes, within us. The "complex" reaction can become so dominant in life that we lose our ability to deal effectively with the perceived loss, whether real or imagined. The means by which our psyches unconsciously attempt to modulate or alleviate our rising anxiety and overwhelming emotions is through defense mechanisms. If successful, we regain the ability to engage with our world and change course if necessary. However, there is a shadow side to defenses; they may actually become so pervasive in our attempt to avoid anxiety that they keep us locked in crippling habitual patterns of behavior.

Defenses are a product of evolution; they provide us with the physical and psychological means to protect ourselves against perceived or anticipated threats. The fight-or-flight response is perhaps the oldest coping mechanism in the psyche. Defenses evolved and developed for good reasons. As we consciously evolved, we needed more options than merely instinctual fight or flight to adapt to our environment. Keeping a society together requires developing the capabilities to deal with the increasing complexity of interpersonal interactions. Defense mechanisms moderate these interactions. They also prevent individuals and groups from being overwhelmed by anxiety. It is only when defenses keep us from adapting to new conditions that they become a problem and prevent us from making positive changes to address serious issues. The defenses that make us resist the reality of disturbing news, such as a troubling medical report on our health or the news of climate change, also prevent us from changing our behaviors to deal with these problems. In these cases, defenses are ultimately counterproductive, harming our well-being and that of others.

We have many ways to defend against anxiety and the experience of overwhelming emotions. Freud was the first to note the various forms of our defenses, and over many years of working with clients I have witnessed a number of them. We certainly note them in ourselves and in our interactions with others. A few of the most prevalent defenses include denial, rationalization, compartmentalization, distortion, dissociation, and projection. How do each of these affect our ability to deal effectively with challenging issues like climate change?

Denial is deeply rooted and is difficult to overcome. In the face of the most convincing evidence we may turn away from the facts. Dissonance arises between what actually is and what we perceive or feel. Presented with disturbing information or news, our initial reaction is often "I just can't believe it" or "I can't accept this, I can't believe this is happening." Denial is not just a conscious process; it may also operate unconsciously, whereby we are unaware of our intense state of resistance to disturbing news until someone points it out. Denial exists to maintain the status quo, a situation clearly evident with the issue of global warming, in which the scientific evidence that humans are warming the planet is overwhelming. Still, many turn away from these scientific facts and live in a world of denial. I have often heard "I can't believe this" after presenting the facts of climate change. These facts create tremendous anxiety and are so intense that psychologically the only way some deal with the emotions is to deny reality.

Others may deal with the anxiety of disturbing news by rationalizing it, by inventing reasons to explain away our disturbing experiences. "It isn't that bad" is often heard when, in reality, the situation really *is* that bad. This defense, coupled with our ability to intellectualize, leads to ill-conceived explanations for climate change, such as "It's really all because of the Sun." In fact, this can't be. Others may state, "There is warming, but it is just a natural cycle," "The effects are overblown," or "More carbon dioxide in the atmosphere is actually good for plants." There are those who do accept the scientific facts but then argue, "We really can't do anything to stop the warming."

We compartmentalize by keeping conflicting ideas or beliefs separate, allowing us to live in a state of cognitive dissonance. This often takes place in cases of trauma, in which the memories of past trauma are kept separate from the feelings associated with the painful event. With regard to climate change, I see this defense present in scientists who are knowledgeable about the facts yet hold a belief or value system in direct conflict with them. They package away these two parts of themselves so they can go on with everyday life.

Distortion involves bending the facts to conform to acceptable beliefs or values or to make the disturbing facts palatable. This is our ability to "see the world through rose-colored glasses" despite the reality of the situation. Someone under the pervasive sway of this defense lives in his or her own world. Distortion is a common means of dealing with global warming's anticipated losses, and some consciously distort the facts to confuse the public on the issue.

Projection is an important unconscious process because it strongly affects how we relate to the world. Certain aspects of ourselves, at odds with our conscious views, are projected out onto others. We project onto things as well as people, and our projections create a veil that effectively separates us from our environment. For example, I once spoke to representatives of the electric-power industry. This is an industry that depends mostly on coal for generating electricity. Their response to my presentation on the effects of increasing greenhouse gases was to point out that they were serving their costumers, who wanted energy for all of their home appliances. They were projecting the problem completely onto the consumer and absolving themselves of any role in the problem.

Finally, dissociation occurs when we reach a point where the pain of the trauma is too much. Clinically, this defense appears in those who have suffered severe abuse, but all of us can dissociate around issues that are disturbing. While giving a presentation on global warming, I stopped to ask how people were feeling. One woman stated, "You know, in the middle of your talk I just spaced out; I can't remember what you said after that." My presentation was so disturbing to her that she dissociated in order to deal with the overwhelming emotions welling up.

Defense mechanisms exist to regulate the amount of anxiety we experience in a situation, but if the defense mechanisms become too strong, we become disabled and lose the ability to deal effectively with the causes of our anxiety. Our defense has turned against us. Again, all of this occurs in the unconscious; were it happening consciously, we would be able to stop the process.

We are faced with the disturbing facts of climate change. The facts ignite fears of loss, personal and collective, and defenses come to our rescue. If they are too reactive, we end up in denial about the issue or choose to distort the facts. However, the facts still exist. The climate continues to change. We need to break this complex cycle to create our flourishing future.

How do we work to break the perpetual cycle of fear-complex-defense? We must address it when the feeling of fear first arises. Often we experience fear and anxiety as overwhelming, but in sharing our experience with others we can tolerate these overwhelming feelings. This is the basis for any healing process. I believe this process has its roots in our earliest beginnings as communal beings. Sitting around the fire and telling our stories is an effective way to heal wounds and envision a future. So one step in breaking this cycle of fear-complex-defense is to be mindful of our feelings and share them with others. After I gave a presentation and asked how people felt, a young woman stated how helpless she felt. When others began to share a similar feeling, the woman felt relief. She told us that she had held this feeling in because she believed she was the only one who felt this way. Perhaps social media is serving this purpose for the younger generations. Though lacking the contribution of direct physical presence, these media do allow for a sharing of feelings. To date much of the communal exchange on climate change has focused on thoughts, ideas, and disagreements. It is time to broaden our sharing to feelings.

Another step in dealing with the fear is to ask what old scripts are being triggered in us when we sense fear or anxiety around the issue; this requires us to be more attuned to what is going on within. We can train ourselves to become more aware of our feelings, which includes how our body is reacting. We pick up the thread with the first stirring and follow it: When did I feel this way before? What old story does this remind me of? Have I been here before? An image may come to mind or, more often, a sensation in the body, which is the first sign of the activated complex, the old script you know so well—even if "know" only unconsciously. Bringing our scripts into conscious awareness can be transformative. We now understand this old familiar pattern and can see how it plays out. This is the power of awareness. We can then look for these old stories in others and in everyday events. Can you detect how these stories are showing up in the world? Our "growth complex" states that we must continually increase power and wealth. It is at the core of the climate-change issue. Asking where this complex appears in us and in the culture opens us to transformation. We become conscious of a strong force within our psyche.

Finally, looking at how we are reacting to our anxiety or fear sheds light on what defenses are active. Do we find ourselves questioning the very existence of the problem? Do we prefer an alternative explanation to the more accurate (but more upsetting) one? Do we place the cause of the problem on someone else? By asking these simple questions we begin to understand how we are responding defensively to fear-inducing factors. This takes some level of conscious awareness, but we need to ask these questions. If we are willing to do so, we reap the bounty of understanding why we behave the way we do.

In that church, opening our hearts to how we felt about the disturbing facts of climate change allowed us to experience our connection to the issue and to one another. We took the first step in transformation by sharing openly our feelings about climate change. We listened to one another's stories. In another gathering, an older gentleman talked of his passion for cross-country skiing. Since his youth this is what he looked forward to every winter. He had grown up in the Pacific Northwest and had his favorite locations to ski. Now, however, because of changes in the climate, he could no longer fulfill his passion in these places. He said he now travels to northwestern Colorado to find good skiing, but recently these places have lacked sufficient snowfall as well. As he told his story, I felt the tremendous loss this man was experiencing. He was living testimony that the climate is changing and how it affects all of us. Here was a man who was losing the thing that had given him joy throughout his long life. How many others can tell such stories?

I feel we need to share more of our stories about how the changing world affects us. In telling them, we make contact with the feelings and anxieties surrounding what we are losing in our changing world, and we begin to recognize how we are defending ourselves against these changes. In ancient times we would have gathered into small groups to share stories and feelings, but those were simpler times. We have grown to a tribe of over seven billion. We live in cities populated by millions. How do we share our stories given this global lifestyle? How do we feel the effects of the world on individuals and communities? The creation of a flourishing future depends on our ability to connect with one another. I believe we can do this. Imagine more of us working toward an awareness of our feelings, our old scripts, and our defenses against our felt worlds. Imagine using our wonderful technologies to connect us in new and interesting ways. We are at the threshold of transformation. We need to step across it to create a better world.

II

Patterns

How Images Facilitate Transformation

When you see the earth from the moon, you don't see any
divisions there of nations or states. This might be the symbol, really, for the new mythology to come.

—Joseph Campbell

One of the most iconic photographs in history is that of Earth
suspended in the darkness of space. Taken in 1968 by the
Apollo 8 astronauts, this single image of Earth rising above the
Moon's horizon makes us aware of the beauty and wholeness
of our planet. Looking at this image, I become aware of the
rich complexity of earth, ocean, air, and ice. Swirling patterns
of bright, shining clouds encircle the world like white cloth–
enshrouded dancers. Beneath these shimmering, irregular
objects rests the vast royal-blue oceans, and the continents
present themselves without political borders. This image of
the planet surrounded by the vastness of space unveils the
natural beauty of Earth.

Why is this particular picture of Earth so moving? I believe that something inside us resonates with this subtle, silent
sphere suspended in space. Curiously, certain images seem to
move us emotionally; they remain with us for life. We may
even tell stories of these images and the way they move us. I
think of Galileo as he stared at Earth's moon, how in that moment his telescope brought that distant silvery object into his

earthbound human experience. Gazing further into space, he viewed the moons circling Jupiter and brought forth to consciousness a revolution in how we view the cosmos. *Images transform consciousness.* They create revolutions in psyche. With Galileo's gaze into the cosmos, humanity was released from its earthbound condition. Our imaginations soared into space, launching us into an age of unprecedented scientific discovery.

Immersing ourselves in the realm of image invites us to imagine. We begin to picture what could be by releasing ourselves from the constraints of what is. When we enter the world of imagination we connect to things unseen. Out of this playfulness emerges the *not yet*. Our imagination incubates the future: in this way we build bridges from the known to the yet to be known. Thus, imagining is the first step to creating a flourishing future.

One of the challenges in communicating the seriousness of climate change is that we have yet to find a universal image connecting people to the issue. The image of the lone polar bear on a piece of floating sea ice touches some but does not resonate with others. A single image may not be possible to represent the diverse implications of climate change. We may need a collection of images that touch specific geographic regions or cultures. Historically, images have been socially transformative. The profound emotional effect of such images indicates their archetypal nature. Of course, the image need not be visual. Consider the role of music, for example, the importance of rock and roll on society during the 1950s and 1960s. Stories have had a profound effect on us throughout history as well; consider works such as: the Bible, Homer's *Odyssey*, Dante's *Divine Comedy*, the works of Shakespeare, or even *Uncle Tom's Cabin*.

In my lifetime, I can point to certain images that played a critical role in social transformation. In terms of visual images, many people remember the picture taken during the Vietnam War of the young girl, her naked body scarred by napalm fire, running down a village road. Others remember the image of Martin Luther King Jr. standing on the steps of the Lincoln Memorial, the clock used by the Union of Concerned Scientists to alert us to the threat of nuclear destruction, or the image of Neil Armstrong taking that first step on the surface of the moon.

The picture of Earth suspended in space became an icon for many. It was more than just a picture; it had symbolic meaning. We realized the beauty and wholeness of Earth, and soon after the appearance of this picture the environmental movement began. Our perception of the world is influenced by image and story. Image and story contain symbols of transformation.

How have our symbols and stories transformed over time? In the Western world, our earliest stories involved the relationship between man and nature, a world where animal spoke to human and human spoke to animal, a world where everything was animate. This prescientific world developed narratives—first oral, then written—providing individuals and whole societies with a related sense of place and purpose in the cosmos. These grand stories evolved and flourished, giving mankind a vibrant connection to being in the world. Inherent in these stories were symbols of meaning. Animals, such as the bear, took on specific significance. They became messengers between human and nature. Bones and stones were carved with intricate patterns holding special meanings. Unique images became archetypal patterns that resonated with both individual and whole societies and provided not only understanding but also a numinous experience of the world. Over time, the narratives moved from interpreting our direct experience with the natural world at hand, to the heavens above, then to a distant divinity residing in the heavens. The early Greeks were perhaps the first to experience the separation between humankind and the natural world. We reached a point at which we began to define our relationship with the surrounding world through a divine grand design. Each of these stories moved us further from a direct experience with the nonhuman world.

The scientific revolution defined a new path for interpreting the world. We developed ways to observe both the vastness of the heavens and the microscopic details of the infinitesimal. We applied our rational acumen to these observations and used the symbolic language of science to bring order and predictability to the observed world. We categorized, classified, and quantified all that could be perceived. This approach assumed that the world was completely separate from us. We adopted this belief so that our measurements would not be influenced by our presence, and this scientific mindset led to an astounding array of accomplishments. We could describe and predict the movements of the solar system and, later, the universe. We could look into the structure of life itself and describe how it evolved through time; with the dawn and development of the Scientific Revolution, reason became our dominant worldview. We also developed the worldview that all that exists is material and can be reduced to separate pieces. The ordering principle became reductionist and materialistic. The symbols representing this worldview first showed God as the master architect, followed by the beautiful scientific instruments of the seventeenth and eighteenth centuries, and now the spectacular images from the Hubble space telescope.

Stories of our separation from the natural world are not new. I am reminded of an ancient Greek myth, an allegory, well over two thousand years old, about a man who wants to build the largest house in town. To obtain the wood needed for his project, he cuts down the largest tree in the sacred forest. He does this despite dire warnings from Demeter, the goddess of the tree, who proclaims that he will suffer grave consequences if he proceeds in his efforts. The man disregards the warning and aggressively fells the great tree, which drips blood upon feeling the blade of the ax. For cutting down the tree the goddess asks her sister Famine to provide a fitting punishment. One evening, Famine enters the man's house and breathes into him a sense of never-ending hunger. In that moment he begins to dream of being hungry. Upon waking he cannot find sufficient food to satisfy his hunger. Eventually, after selling everything in his house, he sells his only daughter to buy more food. Still he cannot end his longing for fulfillment. Finally, the man consumes his own body to end his hunger. This disturbing myth arose around the time the Greeks were building their massive array of ships to expand their might into the Mediterranean. To build such an armada required cutting down most of the trees in the Greek peninsula. The result of this deforestation was the loss of valuable topsoil necessary for agriculture. Thus the story is an allegory about neglecting and destroying nature and the consequences of such a path. It is also a story of how our separateness from the sacredness of nature leads to an insatiable desire to consume our world. The man, whose name translated from the Greek is Earth-Destroyer, was vain and profane. He was cut off from the sacredness of the world around him. This state of isolation led to his willingness to cut down the sacred tree. There is also the arresting image of the father selling his daughter to assuage his hunger. Metaphorically, this depicts how by feeding our current endless hunger we discount the future for our children. In addition, his disconnection from the feminine, also represented by the sale of his daughter, means he cannot relate in a healthy way to others. This tale from the beginning of Western civilization shows how old our tendency is to objectify the world and to feel cut off from others.

A more recent story reflecting this attitude is from J. R. R. Tolkien's *The Lord of the Rings*. We see how the great wizard Saruman the White begins his stay in Middle-earth by living in harmony with nature. He wanders the beautiful forests of this land making friends with many beings. However, Saruman becomes possessed with the desire for more and more power, corrupting his heart and mind. He devalues the trees and other beings in the forests and orders the forests cut down to fuel the fires needed to create weapons of mass destruction. Ultimately, Saruman's lust for power leads to his own destruction. This twentieth-century saga—no doubt influenced by Tolkien's experiences in the Great War—shows how perception rooted in separation and power leads to destruction.

These two stories illustrate the consequences of what happens when we remove ourselves from direct contact with the natural world. The good news is that we can create stories that allow us to see how connected we are to one another and the natural world.

Perhaps the most powerful narrative motif is the hero or heroine's journey. This motif, popularized by Joseph Campbell, describes how one is called to go on a journey, which is awakened through a disturbance in life. The first inclination is to ignore the call, but after overcoming doubt, the hero begins the journey. Soon after, they must descend into a dark place, often portrayed as a forest or cave. Upon entering the dark place and overcoming obstacles, the hero discovers a treasure. Retrieving this boon, the hero returns to share the treasure with the world. This powerful narrative is the foundation of many myths, legends, and tales of old as well as of many modern films. Psychologically, this archetypal pattern is about confronting a problem that cannot be addressed from our dominant perspective. Our tried-and-true way of seeing and dealing with the world breaks down. Our dominant paradigm no longer works. In tales, the image of the ailing king represents an old and worn-out ruling principle. The kingdom does not thrive and has become a wasteland. Because of this, we are challenged to find a new viewpoint, one that allows us to deal with the problem confronting us. In terms of stories, the old king must be replaced by a new one, representing a new ruling principle. When this happens, the kingdom is restored to health. The accomplishment of this transformation in viewpoint often requires deep inner work in which we reflect on what is keeping us stuck or blocked. Therapists accompany us on this inner journey to work with these blocks, or complexes. Very often the inner discovery appears in the form of a symbol, which could be an image, a felt experience, or a new storyline. With this new discovery we can enter our life with renewed energy and the ability to deal with the problem.

The hero's journey not only pertains to an individual's journey through life. It can also represent the trajectory of a whole society. I see this motif playing out around the issue of climate change. The call to begin the journey is the scientific evidence that the climate is changing. The rise in global temperatures, the melting of the ice sheets, and the rise in sea level are all signs that disrupt our inner and outer worlds. The hero often resists heeding the call, which can be seen in attempts both to dismiss or ignore the scientific evidence. Those who heed the call and have awakened to the evidence have begun the journey of transformation. Each person within society, in his or her own way, begins to follow a path to address this issue. Scientists continue on their journey to explore the depths of nature and understand better the direct and indirect consequences of global warming. Engineers work on better ways of creating energy from renewable resources. Policy makers strive to find acceptable political solutions to the issue. Citizens take to the streets to raise consciousness around the issue. Many have begun the journey to finding the treasure of a flourishing future. The descent is hard work, and obstacles are encountered along the path. Merchants of doubt work to sow seeds of confusion around the science, funding is diverted from those working to create alternative energy sources, and politicians refuse to understand the science or consider revenue-neutral ways to address the issue. The descent into the darkness is never pleasant, but it seems to be a necessary part of the hero's journey. Ultimately, the hero discovers the treasure in the midst of this darkness. What is the treasure associated with the hero's climate journey? What boon awaits us within the darkness surrounding the challenges of climate change? From a scientific perspective, the treasure will be a clearer understanding of the climate system. Improved technology will add to our ability to observe what is happening in the world. The capability of climate models to look at smaller geographic regions will improve, and more interactions within the climate system will be included, providing a more integrated sense of what is changing. From the energy perspective, the treasures will be renewable means to generate, store, and distribute energy. On the political side, the treasures will be the passage of legislation that insures a transition off of fossil fuels without damaging the economy. From a lifestyle perspective, the boon lies in learning to live within our means.

It is clear that the new story—our new map for the journey—needs to contain the power and beauty of science but also ways to bring us into closer heart-connection with the world. In this way, our worldview preserves the gifts of science but includes and restores what our early ancestors held in terms of relatedness to the nonhuman world. In essence, our story for a flourishing future holds both head and heart.

In my work on communicating the issue of climate change, I have come to value my interactions with artists. Given the power of imagery and story, is it at all surprising that the disciplines of the arts are essential? During a discussion with an artist friend on the relationships between the sciences and the arts for a group of interested citizens, my friend and I realized the many ways these two fields complement each other. The sciences and arts dance well together. They provide a more holistic and creative way to look at the issue of climate change. Those listening to our conversation became very animated by the possibilities unleashed by this dance. Perhaps this is to be expected, given that artists thrive in the world of image, sound, body, and metaphor. It is truly heartening to see so many artists joining the journey of transformation. *Imagine* what this community will bring back from their hero's journey!

Perhaps it is appropriate for one more story, one more hero's journey, from the twentieth century. Many of the qualities needed to create a promising future are nicely portrayed by the hobbits in *The Lord of the Rings*. The characters Frodo and Sam care deeply about the earth, each other, and life in general. Despite their reserved nature, they are willing to take risks and make sacrifices to work for the good of Middle-earth. These are beings attuned to their place and time. Sam holds the caring sense of the gardener, one who tends to life. He is able to extend this caring sense beyond his personal sphere to the many. Frodo takes on a burden requiring terrible personal sacrifice in order to free Middle-earth from darkness. He consciously chooses to work for the good of all, knowing he will need to overcome great challenges. Tolkien's message in this story is that unbridled power is not the way to create a flourishing future. The way forward is found in the devotion of a humble fellowship working for good, in which each member of their group brings something unique to their collective quest. We can view this tale as a metaphor for how to act toward one another and the greater world. *The Lord of the Rings* is a perfect example of the hero's journey. I feel the overarching theme of fundamental goodness is why *The Lord of the Rings* books continue to be some of the best-selling books ever published. The great popularity of the books—and now movies—indicates that the images and metaphors in the story touch people deeply. Such overwhelming popularity indicates the powerful force of living archetypal images in the story. We are captured by these stories because their images live within us. Psychologically, Middle-earth can be viewed as an image of our psyche, containing our creative and destructive tendencies. Which aspect of this inner world is most prevalent within us? Will we choose the path of blind power, or will we follow the path of connected cooperation? On our personal hero's journey to address the threat of climate change, will we heed the call to face climate change and be willing to overcome the dark forces opposing our progress to a flourishing future?

I believe this is possible, for once a society is seized by a powerful symbol, change happens quickly. This is actually a reassuring fact when we consider the pressing need for social transformation concerning global warming—and the need to act now to avoid the worst consequences of climate change. If the time is right and a unifying symbol strong enough, change will happen. *Given the power of symbols, it is important for us to be conscious of what images, patterns, or metaphors could activate such transformation.*

In the midst of a catastrophe like a flood, strangers come together to help one another. Frequently the human spirit transcends separation and opens to selfless cooperation. The image of many hands working as one comes to my mind. We grasp situations and hold onto what matters. We have the capacity to lend a hand and to take action. We tend to forget these aspects of our being in the world, but they are always present. Reaching out to another in the moment, the other responds, taking our hand. Moving beyond struggle, we tap into our inherent ability to help not only in times of crises but also in times of relative calm.

Imagine weaving together head and heart into a way of perceiving our world, a new worldview that might just lead to a very different way of living. Through such a lens we would view problems like global warming very differently. Our connection to the world would obviate the need to minimize our disruption. We would look for ways to live that would leave a smaller footprint. We would build a connecting web to distribute energy around the world efficiently. Our sense of cooperation would allow us to see this distribution of energy and other resources as a natural part of living on the planet. We would not horde energy but would make sure that it went to those in need, and we would consider the effect of creating energy on biological, physical, social, and psychological aspects of the world.

We must imagine and envision a new way forward. Allowing ourselves to imagine something differently is a necessary step in creating a new way. It means looking at the price we have paid for our old way of living following the old worn-out paradigm of separation. The first step of a new journey is always difficult, but today we have no choice. To build a flourishing world for our children, we must change the way we see things. That iconic image of Earth taken so many years ago was perhaps the first step. Seeing the world without borders and political differences creates an atmosphere for cooperation and connectedness. Seeing how Earth's air, water, ice, and land interconnect and flow together is attunement on the grand scale.

What of the next step on our journey? The world seems full of tension and polarity. People are at odds with one another. How do we creatively hold the tension of opposites?

Opposites and Our Relationship to Climate Change

Everything requires for its existence its own opposite, or
else it fades into nothingness.

—C. G. Jung

I awake early in the morning. It's dark outside, and I hesitate
to get out of bed, but early mornings are a good time for me,
and my lifelong practice of early rising has allowed me to witness the recurring transition from night to day. Standing at
my window, I watch the night sky give way to the first small,
reddish luminescence of sunshine. I continue watching as it
displays a range of reds, violets, and glowing pink hues. Finally, the whole bright white disk of the sun rests above the
horizon. Watching this daily development I experience the opposites of light and dark. For billions of years Earth's turning
on its axis has created this periodic phenomenon. I picture
our hominid ancestors standing on the savannah of southern
Africa watching this recurring event. I imagine their nocturnal fears of predators diminishing with the slow appearance
of first light. Is it any wonder that indigenous peoples created
ceremonies to honor the rising sun, one of our earliest lived
experiences of opposites?

Along with light and dark, sun and moon, our parents form
a primal pair of opposites whose union gives us life. Given

these primal experiences of polarity, it is not surprising that the pattern of opposites has played a critical role in how we perceive and experience our worlds. It is imperative that we explore the role opposites play in our lives; they pervade our everyday world and often give rise to tensions. What are the opposites that we experience? How do they affect our relationship to the world?

Opposites hold polarity by their very nature; you cannot have one pole without the other. Given this fundamental situation, opposites must be a part of our new story for the future. Neglecting the existence of opposites in our outer and inner worlds leads to turmoil; we become too one-sided if we turn away from them. One-sidedness is the seed of discord, for we fail to see the other side of a situation. We do not allow ourselves to contemplate what it is like from the other's perspective. We also fail to recognize that our side depends on its opposite. Day would mean nothing without night, light without dark. If I put opposites in relation, I avoid being one-sided.

A scientist has to get used to defending his or her ideas. Scientists are trained to argue and critique one another's work. We thrive on challenging each other through debate at conferences or by reviewing each other's manuscripts. At times, the disagreements can become very personal, and it may take a long time to settle on the idea that best explains an observation. Nevertheless, the process of determining the best explanation is rooted in facts, not opinions. Scientists naively thought the same process would apply in public debates. We were mistaken about this. The climate debate in the United States and a few other countries is highly irrational. Scientists quickly learned that using reason and presenting facts were unlikely to convince those opposed to the science of climate change. In addition, the disagreements were not about discussing differing ideas but were extremely personal. The moral character of scientists was questioned. Death threats were made against them and their family members. This was a completely unanticipated form of opposition. There is tremendous polarity over this issue. Some accept the science of climate change; others reject it. Some are attracted to building a sustainable world; others are repulsed by the idea. I have encountered these opposites when making public presentations on the issue. At times I would need to stand before a hostile audience and hold the tension of opposites in the room. I could feel strong emotions within and needed to stay calm. I came to realize that the polarity surrounding this issue is deep seated, hence the need to explore patterns of opposites.

Every day we are faced with decisions involving opposites. The decisions range from the simple to the complex. Most of the time we are able to make a decision and move on. At other times, the decision is painful and difficult. I have experienced this process in my own life. When deciding to become a Jungian analyst I suffered tremendous internal tension. Should I invest all this time and money to become an analyst, or should I not? In the end the pull to become an analyst won out. As an analyst, my work with people often begins with identifying an internal conflict that keeps a person stuck. If the internal conflict becomes too intense, it may make life unbearable. Should I stay in this job, or should I leave it? Do I love this person or not? Should I buy that car or not? Opposites can also be at the root of disagreement with others. If your views are opposed to my views, we are ripe for conflict. Think of the last argument you had with someone. How was it affected by the polarity of opposing viewpoints?

On an archetypal level, opposites help us make sense of our inner and outer worlds. Consider how our being here in the world immediately implies a beginning and an end, which defines life. We look around and see newborn babies and the frailty of age at the same time, and we recognize that each of us carries these polarities. Life's process of beginning and ending extends throughout the universe. When we look into the cosmos we see whole galaxies experiencing a birth process while others suffer their demise. Life is an eternal process of coming and going. Knowing our life will end gives birth to our most basic existential anxiety. Addressing this basic anxiety was perhaps the origin of our oldest stories: our myths. Creation stories brought understanding to why things come into being. Stories of the world's end instill fear and hope at the same time, fear that this bodily existence will end and hope that something exists afterward. Seeing these stories as concrete realities often sows the seeds of discord. Many of the conflicts today are rooted in concretized myths concerning creation and destruction. I wonder to what extent our fear over climate change is rooted in our basic anxiety about endings.

Of course opposites also help us in orienting ourselves in the world. They create contrast, which is essential to experiencing the phenomenal world: up/down, north/south, and east/west. Our senses are immersed in opposites. We hear loud and soft sounds, taste bitter and sweet, touch smooth and rough, cold and hot, see white and black. Even our sense of time dwells in the realm of opposites: we look back at the past and peer into the future. Opposites permeate not only our outer world of sense perceptions but also our inner world. We define our sense of self at any moment in terms of moods: happy or sad, agitated or peaceful. We may feel closeness or distance.

Typically, we have a preferred way of orienting ourselves in the world based on one of the qualities of thinking, feeling, sensation, or intuition. There may be other ways of relating to our world, but these four are basic. If our tendency is to use thinking as our way of relating, we are inclined to define the world in terms of concepts and ideas. We are adept at defining and categorizing things in our world, often in terms of abstraction. We focus on knowing what a thing is and how it works. Feeling allows us to see the value of something. In this way, feeling brings us closer to the world. We are less inclined to define something in terms of abstract concepts. We experience a different sense of "knowing," one that involves empathy. We are emotionally moved by the world.

As a scientist my dominant mode of understanding the world is through concepts and thoughts. Scientists formulate ideas about the world and then test them by observing whether the world fits with these ideas. If the concepts do not describe the observations, they are improved through application of the fundamental laws of nature. Science is a constant process of narrowing uncertainty to gain improved understanding. It is not static. Imagine a scientist conversing with someone whose dominant approach to matters is through feeling. The scientist will argue from a position rooted in the laws of nature, but the other person will be listening for the value in what is being said. We are set for disagreement if the scientist cannot speak about value or the other person cannot grasp concepts. The two are at opposite ends of the spectrum in terms of communication. This is often the situation when scientists communicate with the public. They are speaking from the position of concepts while some listeners are hearing through the filter of values, which is usually tied strongly to emotions.

In using our senses we collect information about what is immediately before us. Although the information from our senses is often viewed as the way of orientation that is most reliable, all sense information is filtered; it is a process tinting how we see the world based upon past experiences and values. Two people seeing the same object perceive it differently. Scientists use sensed observations to study nature. By collecting and compiling this information they build a picture of how the world is. Combining observations with thought leads to a more holistic view of the world.

In terms of intuition, we allow our "gut feelings" to influence strongly what to do in a situation. This approach is for the most part unconscious and taps into older parts of the brain that rely on image and metaphor to perceive the world. This particular decision-making process often raises concerns because of its unpredictable nature. However, studies indicate that this approach often provides very effective solutions to problems. The unconscious has developed the ability to assimilate information in an associative way that is much faster than our conscious reasoning. This intuitive way of seeing problems and quickly making a choice clearly has evolutionary advantages. Scientists and artists can be very intuitive in their work. Some of my most exciting insights as a scientist have occurred through intuition.

As a scientist, one of the first problems I studied in my career was Earth's early climate. Carl Sagan pointed out in 1972 that solar physics tells us that billions of years ago our sun was much fainter than it is now. Yet geologic evidence indicates the Earth had fluid water on the surface at this time. Sagan deduced that with such a faint sun, all the water on Earth should have been frozen. How could there be liquid water with so little sunlight reaching Earth's surface? This became known as the Faint Young Sun paradox. Sagan and a colleague argued that perhaps greenhouse gases could resolve the paradox. He speculated that abundant amounts of ammonia would have acted as an effective greenhouse gas to keep the planet above freezing. Scientists later argued that ammonia could never have built up in high enough concentrations for this to work. It was then proposed that extremely high levels of the greenhouse gas carbon dioxide kept the planet above freezing, but the levels necessary would have been too high to agree with what the geologic record told us about Earth's early atmosphere. What to do? It was around this time that I began to think about the paradox. I had an intuition that early life could play an important role in solving the paradox. At this time, I was reading about Earth's earliest life forms, which were bacteria. These small organisms existed without oxygen and released methane into the atmosphere. I was able to show that by accounting for methane in the early atmosphere one would not need as much carbon dioxide to keep the planet warm. Methane is a very potent greenhouse gas, over twenty times more potent than carbon dioxide, so a little bit of it in the atmosphere can go a long way. By assuming Earth's early atmosphere was composed of a combination of these two greenhouse gases one could solve the paradox. Since my work was published, many more embellishments have been added to solving this paradox. My intuition about the connections between early life and its effects on the climate proved to be quite fruitful. My scientific career has been rich in both serendipity and intuition. In talking with colleagues over the years, I have heard many stories about the important role intuition and serendipity have played in their scientific discoveries.

As a Jungian analyst, intuition plays a critical role in how I work with clients. Sitting across from someone who is suffering, I often will have an intuitive sense of what the person is feeling or thinking. I often check in with the person to see if my intuition matches what is going on within the person. This resonance between the two of us creates a strong therapeutic bond that aids the healing process tremendously. I have even experienced instances during initial meetings with a prospective client of what the presenting problem is before they tell me. Once I greeted a young man who had wanted to receive career counseling. When I shook the man's hand I knew immediately he was actually depressed, even though he outwardly looked fine. After a couple of sessions, the man began to talk about his depression, and that was why he really wanted to see someone.

Sensation and intuition are opposites; however, it is important to understand they are not mutually exclusive. If we have a tendency toward one of them, it does not mean we lack the other. Similarly, thinking does not imply a complete absence of feeling, only a tendency to depend on it more often.

These opposites make themselves felt in how we make decisions about our world. We may choose to look at things rationally, in which we collect as much factual information as possible and use reason to order and interpret this information. We then weigh options and make the "best" decision possible given what we know. This rational approach appeals to some because it attempts to be thoroughly objective, unaffected by any emotional relationship to the information gathered. Many believe this is the way all of our decisions should be made, especially political decisions, which have wide-reaching effects, yet we know from experience that most decisions are not based solely on reason.

Becoming too one-sided in our approach to seeing the world limits our ability to find comprehensive solutions to problems. It is like a marriage in which one of the partners is very logical while the other approaches life through feelings. These fundamental differences in seeing the world often lead to tensions in a marriage. However, if the couple learns to recognize and value the other's perspective, they can make far more effective, agreeable decisions. The key to this balanced way of living is learning to respect the other's perspective. If the head cannot value the ways of the heart, the two will not work together. Similarly, the ways of the heart need to value those of the head. Ultimately, we need to find ways for head and heart to cooperate creatively.

There are other archetypal opposites that strongly affect how we experience the world. Perhaps the most fundamental of these is the masculine/feminine dyad, which pervades all cultures throughout history. It is essential to note that these two qualities are present in both males and females. They transcend gender, and one quality is not better than the other. The masculine qualities usually include discernment, critical analysis, action, and force; feminine qualities include reflection, synthesis, patience, and receptivity. The positive side of the masculine is the ability to assess a situation, strategize to come up with a plan of action, and implement the plan. The positive side of the feminine is to reflect on the whole situation, consider the connections among the various parts of the whole problem, listen to those involved, synthesize everything, and propose a plan that is inclusive. The masculine approach usually is implemented quickly. The negative aspects of the masculine approach is that it may exclude many people involved in the problem, may overlook critical connections embedded in the problem, and may be forced onto those who were not included. The negative side of the feminine is that it can miss critical moments when action is essential and may never reach a desired state of consensus. The dark side of the masculine is that it can be violent and aggressive; the dark side of the feminine is manipulative and destructive. Within each of us are these various aspects of the masculine and feminine.

The best place to see these qualities are in myths, legends, modern fiction, and films. I say the "best" place because we have a harder time identifying them in ourselves, especially the negative and dark aspects. Our stories carry these qualities for us. Indeed, they do this because we project those qualities onto these art forms. I have already mentioned the Greek myth of Earth-Destroyer and *The Lord of the Rings*, two narratives that contain excellent examples of the light and dark masculine and feminine. Interestingly, many stories are concerned with the light qualities overcoming the dark ones. Another motif that is nearly universal is that of the redemption of a feminine form from the clutches of a dark masculine one. Most fairy tales carry this redemption motif: the young daughter being saved from the sadistic Bluebeard, or Luke Skywalker rescuing the princess (his sister) from Darth Vader.

Throughout Western history we see the loss, disregard, and degradation of feminine qualities over those of the masculine. There is no question that the masculine qualities— positive and negative—have dominated the world for centuries, if not millennia. Although both men and women hold these two qualities, men dominate our world and mainly express masculine qualities. This situation has actually become worse over the last decade, as witnessed in the rising abuse of women worldwide. The feminine in almost all forms is missing in our world. We have failed to hold both the masculine and feminine together in a creative fashion. This failure appears in many places. Consider how we have dealt with the environment. Our relationship to nature has been mostly a stance of conquering and subduing it. The historian Carolyn Merchant has shown how our relationship with nature is mirrored in how women have been treated throughout history.

Another archetypal pair of opposites is disunion/union. Disunion, or separation, comes in degrees and cleaves us from the world. When I reflect on how we enter into a state of separation with our world, I realize that it may happen so subtly that we are often quite unaware that it has taken place. One moment I am engaged with my world, listening attentively to the person sitting across from me. In the next moment, I may find myself looking at the other as if through a telescope. I hear them and see them, but the sense of presence has departed. Have I left them, or have they left me? Perhaps we have created an unstated mutual agreement to take a moment's leave. Clearly, however, a separation has taken place. A distinct, disturbing distance has descended upon us, which may last a brief moment or linger with seeming indefiniteness. The feelings associated with this parting vary. Loneliness may arise in that moment, but equally possible is a feeling of release. There is no question that separation is a pregnant possibility in any moment of connectedness.

What price do we pay for such disunion? If we separate too far from our world, we gaze upon it as if it were some distant, lifeless object. It is hard to feel anything so distant. We lose touch with the other, literally. Separation seeps in and kills something in us. I believe the prevalence of disconnection from the felt world is at the root of many of our environmental problems. These days so many spend so little time in touch with the world of nature. We move from home to car to work and back home. Our lives are spent encapsulated within opaque walls. Separation gives birth to emptiness, which we feel compelled to address. Isn't it ironic that our yearning for more arises from our separation and that this separation then feeds the emptiness and sustains our yearning? Separation is

self-supporting and insidious in suspending our connectedness to the world.

Separation makes our world invisible. This sense of invis-
ibility is a part of our inability to connect with the climate
problem. Much of the change occurring today is in the relatively unpopulated polar regions. The melting of sea ice and ice
sheets is not palpable to most. Spatially speaking, the largest
climate changes remain distant. Similarly, the largest changes
will occur still in the future. The problem is largely invisible to
us both in space and time, and until we sense the nearness of
something, we are rarely willing to engage with it. We have a
tendency to wait for change to affect us directly before we are
moved to act. We act reactively rather than proactively. This
tendency is not specific to climate change; we also are reactive
with respect to economic decisions and other social issues. As
long as the problems are distant in space and time, we turn our
gaze away from them. We seem to be wired to react to the immediate but must now transform to react to what lies ahead.
*We are called to make the invisible visible at this great turning
point in civilization.*

In my scientific career I have experienced this process of
making things visible. Research had shown that increases in
the amount of small reflective particles in the atmosphere reduced the amount of sunlight available to warm the climate
system. Essentially, the particles act to shade the planet's
surface. These particles result from increased burning coal,
forests, and other materials. Since an increase in greenhouse
gases warms the planet, the question arose as to how much of
the warming was offset by the increase in these small reflective
particles. A colleague and I carried out research that showed
that the offset was quite significant. On the regional scale,
the smooth geographic pattern of warming from greenhouse

gases was disrupted by the presence of these particles. This research aided our ability to discern and attribute changes in the climate system to humans versus to natural causes. By combining the known effects of greenhouse gases and reflective small particles, we calculated the total effect of human activity on the climate system. This research on the climate role of these small particles continued for a number of years. With each study I made the invisible visible. Science is a continual process of making what is unknown known.

In my career as a Jungian analyst I also make the invisible visible. Unseen complexes in the unconscious can cause tremendous discomfort in a person's life. I have worked with many people who can visibly see their problematic behaviors or thoughts but are unable to find the invisible factors causing their suffering. Working together, the analyst and client are able to unveil the hidden factors that cause the suffering. This therapeutic process of exploring the visible and invisible dimensions of the psyche is tremendously rewarding for both client and analyst. Each learns something about the other and themselves in the process. Jung felt that in this healing both analyst and client are transformed, and I can testify to this.

Looking back on this meditation you may note the frequent appearance of words beginning with "dis-." Opposites create discord, reminding me of other words associated with opposition: disagree, divisive, discount, disassociate, disconnect, and distant. All of these words relate to the issue of climate change. Media outlets sow seeds of divisiveness around the science of climate change. Some economists discount how much the damage caused by a disruptive climate will cost. Listening to all of the bad news around the topic can lead us to disconnect from it. We feel a need to distance ourselves and ignore it. What is this *dis* that pervades our feelings of opposition? Dis is an old Roman name for the king of the underworld, also known as Hades. He was a wealthy ruler who owned what was buried in the ground, which evokes images of the monetary riches tied to fossil fuels. Dis could make himself invisible and enter the world of the living. He was considered both good because of his wealth and bad because he was associated with death. Whenever a disagreement arises, Dis appears seemingly from nowhere. He is always standing in the wings ready to unveil himself and sow discord. There is value in bringing Dis into the discussion! Acknowledging his presence can be the first step toward compromise. In personifying our discord, we step outside of taking sides and gain the perspective of an observer, which is the beginning of agreement. We are willing to see both sides. This is what is so lacking in today's world of Dis.

I have discussed a number of opposites: creation/destruction, masculine/feminine, and disunion/union, which create discord/accord in our lives. How do we learn to deal with opposition? How can we creatively hold and work with opposites? According to Buddhism, our very first step in perceiving the world begins with splitting the world into opposites. We use these opposites to orient ourselves in the world, but the problem then becomes that our perception of these opposites affects how we see and experience the world. We become trapped in the perceived field of opposites. The next step of perception is to value one side of the split opposites over the other. We allow our emotions to take sides, and we thereby become invested in a particular position, which we will defend. Our inner split is now thrust out onto our relations with others who may not hold our position. In addition, we add our own conditioned concepts—our values, beliefs, prejudices— to label what we see in the world. This is what we are seeing in the world today. Democrats and Republicans cannot reach a compromise on any issue, including climate change. The polarity of opposites around the world leads to conflict, war, and destruction. This is a succinct way of understanding how we form opposites and how they lead to tension in our lives. According to Buddhism, it is our ignorance of these processes that leads to the *dis-ease* in our lives.

To work with polarization requires an ability to consider both poles. We consider the emotional investment we have in choosing a particular side, we look at how our value systems may be affecting our choices, and we place ourselves in the position of both sides to weigh and decide how to move forward.

Our journey back to connection with the natural world is transformative work. The movement from a state of separateness to one of wholeness begins in recognizing the existence of both poles of opposites. Working toward this transformation brings balance to life, which is critical to reaching a flourishing future.

Movement to wholeness requires a radical transformation in how we see the world, but I feel such movement is needed to address climate change. We need to balance opposites in addressing this problem, the opposites of thinking with feeling, and sensing with intuition, the balance of outer and inner, masculine and feminine. The integration of opposites is critical to addressing complex issues like climate change.

Balancing the Opposites of Climate Change

Be aware when things are out of balance.

—*Tao Te Ching*

Living on the edge of a somewhat wild environment at the foothills of the Rocky Mountains provides me with daily reminders of life's diversity. Mule deer wandering along the hills munching on shrubs and trees, the occasional sighting of a mountain lion or, at least, the signs of their nocturnal presence, hawks hovering overhead and foxes prowling on the ground, both looking with keen eyes at a lonely field mouse in the low-lying grasses: all paint a portrait of the richness and variety of life in the world. The closer one looks, the more apparent it becomes that these beings rely on one another to sustain a stable environment. No single species ever takes control, for if one species were to dominate an environment, the whole system would eventually collapse. Nature does not tolerate imbalance; it seems to have an innate capacity to correct for such discord. In this regard, we have much to learn from nature in either wild or not-so-wild settings.

Natural environments constantly strive to reach some form of dynamic balance. Such balance is never perfect, and it need not be; all that seems to be necessary is that these environments never drift too far from their inherent state of equilibrium. If there are too many foxes, eventually there are too few mice, which then leads to a decrease in foxes, which continues until a new balance exists between foxes and mice. Of course, ecosystems are far more complex than just foxes and mice, but even with a multitude of species, including grasses, trees, insects, rodents, larger mammals, and birds, the system enters a dynamic balance that exists among all the actors. Ecosystems include the rainfall, moisture, and temperature of the larger surrounding environment. Ecological theaters involve worlds within worlds. Ecosystems are inclusive, enfolding the diversity of multiple worlds. Within these complex environments, a graceful dance of life is at play.

The theater in which all of this plays out need not be a small local environment like the foothills of the Rocky Mountains. Consider Earth's climate, which is defined by a fine balance between the energy we receive from the sun and the energy radiated away from the planet. Left on its own, this flow of energy establishes a dynamic balance insuring a stable climate system. There are certainly natural variations in this system, just as there are in the local ecosystems, but overall these systems are very stable. Even on geologic timescales Earth has kept within a range of temperatures able to support life. Over this long span of time key influences on climate did change, such as the position of continents and the amount of carbon dioxide in the atmosphere. These changes resulted in shifts in Earth's climate, and life responded to these shifts. It is important to recognize that when these and other key factors shift, Earth's climate shifts too, which is a general rule in maintaining balance within any complex system.

Achieving a new balance after a change in conditions does not imply that everything remains the same, just as it was before the change. Think of this in terms of a personal budget. Assume I have a career with a certain stable income. If I am conscientious, then I will develop a lifestyle that assures my personal budget is balanced. I spend about as much as I take in. Now assume a change occurs: my business is in trouble, and I take a pay cut, or I must take a new job that pays less. If I continue to live conscientiously, then I must reduce my spending to match my lower income. I establish a new balance. I am again living within my means, but my lifestyle has had to change. Being in balance after a change in income forces me to live differently than before. This process is true of any system that strives to be in balance. It works at the small scale of my personal income all the way up to Earth's ability to establish a stable climate.

Social systems may also establish some form of equilibrium. Our history is one of developing ways to live together in some semblance of stability. We develop ways to communicate, common rules of behavior, and agreed-upon currency systems to regulate our social environments. There have certainly been large variations along the way, but overall, human civilization has established systems that attempt to provide security and steadiness. The fact that human society has not collapsed, that the world has survived major wars and political upheavals, is a sign that social systems are regulated in some grand way. From an evolutionary perspective we would expect societies to develop ways of self-regulating. As noted, the belief systems of equality and authority each bring benefits to society. Those who adhere to a belief in equality care about the safety of the whole group. They will make decisions ensuring that resources are distributed in a way that helps the many. Those who believe in authority will create systems that establish order within societies. They will focus on securing resources.

The dynamic interplay between these opposing belief systems provides stability to the whole. In Western history, one could argue that the Catholic Church played such a stabilizing role in the Middle Ages. It created an environment that contained knowledge, order, and hope for society. Of course, all elements in a system contain their counterpart, and with the church it was rigidity and autocratic power.

In our personal interactions we regulate how we behave with others. We develop a persona, a predictable outer appearance, in order to interact with the world in a mostly harmonious manner. Internally, we develop a sense of self-identity, a unique personality, which is identifiable by our selves and others we meet. Our inner world is also populated by complexes, which act as independent agents with their own agendas. These autonomous actors can be at odds with our conscious thoughts and actions. They create a vibrant but also often troublesome interior psychological ecosystem. Do these internal actors develop their own form of dynamic balance? Apparently they do, since we possess a relatively stable sense of self that allows us to maneuver through life. Certainly, there are those moments when we "lose it" and are not in balance with the outer world, but most of the time these experiences are transitory, and balance is restored.

Jung argues that the psyche, as a whole, is a self-regulating dynamic system. If an individual's psyche experiences a significant shock or alteration, either from the outer or inner world, then the archetypes provide a restorative force to life. How does this happen? Essentially, the outer disruption activates our inner world, and tension develops between this unconscious world and consciousness. This tension leads to the development of a symbolic image, establishing a new balance between our inner and outer worlds. This regulation requires that we develop a conscious relationship with our inner world, a process requiring us to meditate on the fantasies, dream images, and synchronicities in life. It is a path that requires us to honor both the nonrational and the rational in our lives. Such an approach certainly flies in the face of today's predominantly rational worldview, but I believe that the major issues confronting us today call us to open ourselves to the unseen and disregarded.

What does happen when things get out of balance? Consider an ecosystem experiencing a persistent state of drought. The vegetation that flourished under previously predictable rain patterns begins to disappear. Insects and small mammals that depended on sustenance and protection from this vegetation begin to die off or relocate. Predators that depended on the smaller life forms begin to die off or migrate to more vegetated regions. The whole landscape of the ecosystem, not just a specific type of vegetation, dramatically alters because of the change in rainfall. Eventually, the system falls out of balance. It then seeks to establish a new stable environment, one that will be steady but vastly different from before.

We see a similar phenomenon occurring with our climate system, in which we are causing a vast change to the planet's balance of energy. Burning fossil fuels increases the amount of carbon dioxide in the atmosphere. Since carbon dioxide absorbs energy radiated from Earth's surface, an increase in this gas leads to less energy exiting the planet. This trapped energy upsets the preexisting balance of energy, causing Earth to warm dramatically. This imbalance and its accompanying warming will continue as long as carbon dioxide continues to increase in the atmosphere. To be sure, there are natural ups and downs that exist with this warming, but overall the long-term trend attributable to our activities is causing the world to warm.

How are societies affected by the very changes that they are causing? Dramatic changes to the global environment are leading to social imbalances, and the imbalance in the global environment is sowing seeds of discord in the world. We are seeing the effects of this already: poor nations contributing little to greenhouse warming are suffering increased flooding and rising sea levels. But even the richest nations are beginning to experience social disruption ultimately attributable to climate change. It is estimated that by the end of this century over one thousand major cities in the United States will be at risk from rising sea levels. Coastlines with increased risk of flooding from storm surges will require significant inland migrations of millions of people. Other likely changes include disruptions in food production and the potential spread in infectious disease.

These changes also involve psychological disruption. Demands on mental health centers increased dramatically after Hurricane Katrina hit New Orleans: reported cases of depression tripled after the devastation. Individual psyches strained under the enormity of the change wrought by the hurricane. Our social support systems already struggle to deal with the changes that are taking place. What of the future? We will be challenged to deal with the large disruptions yet to occur, given that society is already showing signs of strain.

The disturbing news of our changing world has led to a growing sense of anxiety and *dis-ease* about our very being in the world. The constant barrage of bad news about the world has pushed us into a state of psychological imbalance. I have noted the ways we protect ourselves from bad news, by using defenses to try to restore some semblance of stability to our lives, but the new balance put in place may not be beneficial. Defenses may create a state that is momentarily stable but not always healthy for the individual or for society at large. They enable us to get by, but not in an optimal and flourishing way.

Our commitment to burning fossil fuels arises from our need for more and more energy. The critical social factors that drive this need are the increasing number of people on the planet, the increasing consumption of energy, and technological innovations that require more energy. The increasing global population places a tremendous burden on natural resources. It also appears that, per person, people are using more energy than in the past. The explosion in consumerism has led to owning more things that require energy. Our overconsumption drives our desire for more energy; until we rein in this desire to consume recklessly we will continue to place our world out of balance.

I have focused on many outerworldly imbalances, but what is the deeper imbalance that leads to the problem of disruption? It is our desire to consume more than we need. What drives this desire? I believe it is a sense of emptiness within. In distancing ourselves from others, including the nonhuman world, we have created a lack within. In not tending to the inner world of psyche, the unconscious, we create a lack within. In losing a sense of the greater meaning of our lives, we create inner emptiness. All these ways of distancing—personal, interpersonal, transpersonal—have led to a yearning for fullness. Our accumulation of things in the world is an unsuccessful attempt to fill this existential emptiness. We have created corporate entities that continually tell us that increased consumption will make us feel better inside. Sadly, we seem to be passing this old paradigm of overconsumption on to the next generation. We need to lift the veil on this false belief and move forward. The continued consumption of the world will not alleviate our sense of emptiness. It is essential that we address this fundamental internal imbalance in order to create a flourishing world.

How do we begin to reestablish balance at these various levels to ensure a positive, creative future for the world? What critical transformations must take place? We often look to organizations to effect change and address imbalances in the world because we tend to feel that these problems are too large for individuals or small groups to do anything about. However, transformation must take place simultaneously at *all* levels. Transformation at the level of the individual is critical to achieving global solutions.

In this spirit, let us start at the level of the individual and work our way up. A critical thread to our new story of flourishing is recognizing that *in order to restore balance to the planet, we need to restore balance within ourselves.* Balance means becoming more complete, more whole, and able to adopt different viewpoints and approaches to solving problems. Without a balanced perspective, we become one-sided and are less resilient to changes that occur in life. One-sidedness locks us into intolerance to other viewpoints, creating cynicism rather than creativity. One-sidedness leads to isolation from the world. Striving for balance is critical if we want to thrive in the world.

Where to begin? I believe a better understanding of the conscious and unconscious aspects of psyche is essential, that is, how unconscious motivations and inner patterns of behavior affect our relationship with one another and the environment. Bringing more consciousness to our relationships with one another and with the environment is critical to creating our new story.

We live in a world that values outer-directed, rapid action. This extraverted attitude has produced a rich material world;

however, the price we pay for this one-sided approach is a loss of reflection on *why* we produce so much. A more balanced approach to living would honor equally the gift of thoughtful, patient reflection on how to face the world. Appreciating a quieter attitude toward our world places us into a more sacred relationship with it. We need to value time spent looking inward along with time spent acting outward.

I believe there is also a need to balance our more concrete view of the world with one rooted in imagination. Our concrete way of seeing the world values well-defined, hard facts. It is a world quantified through the measurement of size, mass, and duration of material objects. Although this perspective is necessary to build and produce our material world—our homes, our farms, our schools—something else is needed to balance out this concrete view. We need to balance our sense of quantity with quality. Imagination lives in the ethereal realm of image and metaphor. It is the font of creativity and is necessary for envisioning a flourishing future. We need to imagine a world without fossil fuels, a world where we tread lightly on the earth and in which food is distributed more equitably. A world where we are living in cooperation, rather than competition, with nature.

Animate awareness is also important, as ours is a history of viewing the world as essentially inanimate. We have drifted far away from seeing a living world, and we view the material world as existing to serve our endless needs and wants. In our journey toward objectivity we have lost an awareness of the value of life. We have departed the animal kingdom for good, despite the fact that we are still very instinctual. Research continues to unveil the intricate social and psychological abilities of all species and also describes the intricate ecological interdependencies between species and environments. It is time we recognize our active participation in a fully animate world. We need to transform our consciousness around the issue of the material world, a transformation bringing psyche and matter together, in which our conscious engagement with the animate world allows us to recognize the animal within.

We are in a reciprocal relationship with the world, what we could call mindfully engaged being in the world. If the awakening I am arguing for occurs, we would no longer be separated from the world to the degree that we are. This marriage of psyche and matter is not an abstract academic issue; it is an important aspect of everyday life. In any moment we can choose to remember to be aware of our sense experiences. We can choose to imagine how our world could be better for all. We can open ourselves to the animate nature and beauty of the world. In such moments we lift the veil that separates us from the world.

Jung felt that all creativity began with the play of the imagination. The play with inner images often leads to great ideas and inventions. Our educational systems stress the analytic over the synthetic, the concrete over the imaginative, action over reflection. What if we designed our schools to value a balanced approach to these qualities in life? Wouldn't we be better off facing the future with access to all of these perspectives? A balanced approach to life makes us more resilient to change, and, ultimately, being balanced within allows us to live in balance with the outer world.

Our new story weaves together analytic thinking and synthetic valuing, extraverted and introverted attentiveness, the concrete with the imaginative, and the masculine with the feminine. Balancing these qualities allows us to see things more comprehensively. Imaginative envisioning is necessary for creating a sustainable, flourishing future.

To live a balanced life requires us to consider how we relate to the world around us. Fundamentally, we need to understand the nature of our being in the world.

III

Being

Exploring Our Being in the World

A mood assails us. It comes neither from
"outside" nor from "inside," but arises out of
being-in-the-world, as a way of such being.

—Martin Heidegger

I walk down a tree-lined street on my way to buy my morning coffee. Suddenly, the melodic singing of a bird interrupts the quiet. Looking up, I see a grey-feathered mockingbird perched atop a telephone pole. As I walk under the pole the bird stops singing and flies to a nearby tall tree. Perhaps my presence has disturbed his morning ritual? Continuing on, I see a kaleidoscopic collection of multihued flowering plants and bushes; virtually every color of the rainbow is present in this small garden. In this moment, the world's presence unveils itself, and I am invited to be an active participant in its being. I feel connected to all that surrounds me. Continuing on, I see my favorite tree, a massive redwood, extending high into the sky. It is so tall that I must arch backward to see its top. While communing with this great giant, the neighborhood policeman approaches with a John Wayne gait. He is a big, burly man with a drooping, grey mustache, looking as if he just stepped off a Western film set. He tells me how he had been walking past this tree for a long time before "discovering" it one day. We both stand quietly before the tree in a state of awe. He bids me farewell, continuing on his morning beat, and I reach out and place my hand on the trunk of the tree, the bark soft and fibrous. The contours of the red, corrugated surface reach out to the shape of my hand. Tree and I are present to the other in a state of "being-with." Lifting my hand from the tree I continue on to my ultimate destination, the warm, welcoming environment of a well-lit cafe.

Many of my meditations in this book focus on our being in the world. How *are* we in this world? How does our being affect the world, and how does the world's being affect us? Although we may be inclined to dismiss such questions as mere "philosophy," they are, in reality, important to our future. Nothing could be more important at this moment in time than the exploration of these questions, for much of our current situation seems directly related to our *not* reflecting on them. As long as we dismiss these questions, our presence in the world will continue to create crises of imbalance.

Exploring these questions illuminates the inextricable interconnectedness between our being and the world's being. This connection is so essential that the twentieth-century German philosopher Martin Heidegger coined the term "being-in-the-world" to describe this creative coparticipatory process.

———————

Our being in the world is a true wonder. Reaching out to touch the world, I am touched by it. This realization means that the idea of a split between us and the world, subject and object, is too limiting at this point in our history. Many have challenged this false dichotomy in Western thought, perhaps no one more persistently than Heidegger. From this perspective, my being is not independent from the world but born from the matrix of the world's presence.

For centuries people have attempted to address the issue of how we—as individuals—are related to the surrounding material world. How are psyche and matter entwined, if at all? Answering this question inevitably leads to how we value the world around us. Reflecting on our being in the world helps us understand why we treat the world as we do.

According to the philosophers Hubert Dreyfus and Sean Kelly, our experience of being in the world has evolved over time, at least in the West. At the time of the early Greeks, when Homer was writing his great epics, the view of our being was more organic. Our existence, as well as nature's, was in a constant state of unfolding, in which we emerged with the world into a state of collective being. In this worldview, our being in the world was directed and informed by the gods, who were present in all aspects of life. It was a time when we were less consciously aware of a difference between world and self, for all was suffused with the gods.

Later, during the time of Socrates, a different view of being in the world arose, one in which our presence played a more creative role. The Greeks believed that things in the world held an inherent purpose and that through our skills as craftsmen we brought out the inherent or hidden nature of things. Psychologically, this stage of development meant that we stood apart from things in the world. This separation from the world gave us the ability to discern differences. The process of creating needs a certain perspective, requiring a stepping away from the thing created. This perception is evident in Plato's description of how the cosmos was created by a divine being. For Plato, the real things of the world are abstract Ideas, which do not dwell here. The world that we see and touch is relativized to the world of abstract Ideas.

The next stage in our development of being arose from the Romans, in which creating things is no longer attributable to our unveiling the forms inherently within matter; it is instead an imposition of our will upon matter. Here, power over the world is preeminent; we establish order in the world through willful force.

The Roman view was followed by the medieval Christian vision of an external creator defining our being, which the church fathers related to the earlier Greek views. The Christian creator making all things in the world imposes his will from afar. Like a potter, God creates the world and imposes order on the cosmos. This view altered our relationship with the world in a radical way: we no longer played a co-creative role in the world. Our place was to tend to the things made by a distant, all-knowing entity. We lived with this view of the cosmos for over one thousand years. Then in the sixteenth century the French philosopher René Descartes placed human reason at the center of the world. In this view, our subjective being is completely separate from the objective world, and this dualistic perspective continues to dominate our understanding of being in the world. Descartes' view serves as the foundation for all modern science, and dualism has been our living story for the past five centuries.

In the twentieth century we entered a new stage of being in the world as a result of the overwhelming presence of technology. Of course, technology has been with us since the dawn of civilization, but we have reached a point at which our being is increasingly defined by the technological devices we create. In fact, many of the devices created by our technologies now strongly affect how we see and experience our world. Essentially, we often find ourselves in some sense in service to the tools we have built to serve us.

It is important to recognize that all of these stages of being are very much alive in the world today. Some people adopt a way of being that places them in a close, sacred relationship with nature; others view their relationship to the world through the lens of their Judeo-Christian beliefs. Many hold on to the dualistic split defined by Descartes; others immerse themselves in modern technology. The tendency is moving more toward the power of technology in defining our being in the world. Look around and you will see people engaged with their cell phones, notepads, laptops, or other electronic devices. We are continually texting or talking to people, making notes, and surfing the web. Phones are not just cell phones but are now called "smart" phones, denoting how we assign "intelligence" to the device at some level. It would be hard to imagine life without our technological gadgetry; with each generation we become more immersed in and dependent upon our technological innovations.

As with all things, however, technology holds opposites. The positive side of our phones and computers is that more people are connected around the world, and we can access nearly infinite amounts of information. We can download and enjoy vast amounts of facts and figures, including great literature and music from all ages. Our being in the world is enriched and enlivened by this connectivity and communication. The shadow sides of our technologies are equally manifold. Our attention is pulled in many different directions. We speak of multitasking as a positive aspect of our lives, yet it fosters an apparent global epidemic of generalized attention deficit disorder. Our ability to focus on a single task with any depth of concentration seems to be fading into history; we jump from one thing to another in milliseconds. Immersed in mountains of data, we find it harder and harder to sort through it, to find information that holds true value for us. Global warming is a good example of this: to the uninformed viewer, blog sites meretriciously denouncing the science of global warming appear as sound as those containing accurate science.

At this point there is no turning back on our technological trajectory. There is increasing emphasis on integrating more technology into our everyday lives. We have eliminated the private self, and our more imaginative technologists are working toward a full integration of machine and human. In achieving this goal, these technologists claim, we will have finally conquered our finitude. The implications of this achievement in terms of our being in the world are staggering. The elimination of our finitude will lead to an increase in population. Where will everyone live? How will they be fed? Will children become obsolete? Will this technological dream lead to the end of humanity?

A sense of distancing comes with many of our technologies. We are caught in a worldwide web of information, in which social media now defines our being in the world. The images and metaphors that constantly stream across the Internet create a new way of being. Often certain images hold tremendous power, suggesting the archetypal nature of the message or image. When an image goes viral, it is similar in nature to the constellation of an archetype in our psyche. Perhaps the phrase "World Wide Web" itself has become a simulacrum of our technological psyche. We use the web to create alternative worlds that assuage our fears and confirm our beliefs. It is hard to escape the web, even as we use it to escape the felt world, which lies just a few steps away from our computer screens.

If our being in the world has indeed been hijacked by technology, how do we find our way back to being in direct touch with the world? To answer this question we need to consider how we *are* in the world. What are the qualities that define our basic being in the world? To be in the world involves a sense of *space* we inhabit, the *time* in which we experience the world, the sense of our *bodies'* presence in the world, and the quality and degree we are *related* to our world. No doubt there are other aspects that could be added to these ways of being in the world, but these four seem essential to how we *are* in the world.

How does our sense of space affect our experience of a changing world? If upsetting changes are distant, then we are prone to adopt an attitude of "out of sight, out of mind." As changes associated with global warming increase, eventually there will be nowhere to hide from its effects. The significant damages from intense storms already foreshadow what is to come. We may try to avoid the reality of climate change by living in increasingly enclosed environments, but the effects will nevertheless become inescapable. Some argue that we will start to address this issue only when the felt damages of global warming finally hit home.

If we were to change our way of experiencing space, we would feel more connected and aware of what is happening in the world. Some argue that this is what technology provides. We use the Internet to "see" all around the world, to "be" in so many different places. The problem with this technologically defined experience of space is that it is a false felt space: it lacks the direct, immediate, connected, tangible sense of space. Indeed, studies have shown that viewing nature on a computer screen fails to provide us with a deep sense of connection to the world. A simpler approach to solving this problem is physically to place ourselves in direct contact with our world. This does not require traveling to distant locations; we only have to become more mindful of the opportunities to connect: a walk through a park or observing the playfulness of an animal.

How does our sense of time affect our experience of a changing world? For an issue like global warming it turns out to be extremely important, for Earth's climate system has a vast memory. Most of the increasing energy trapped by greenhouse gases is stored in the oceans. Just as a pot of boiled water, even after the burner is turned off, holds heat for a long time, these great reservoirs will hold the additional human-caused greenhouse warming for millennia. If we were to stop burning fossil fuels today, Earth's climate would continue to warm for centuries because of this stored energy in the oceans. Unfortunately, this fact is often used by some to argue, "If the problem has such a long timescale, why should we bother to do something about it?" This argument neglects the fact that the longer we wait to reduce carbon emissions the more heat will be trapped in the oceans, committing ourselves to living with greater warming for a longer time. If we wait until the year 2100 to do something, it will take natural processes over one hundred thousand years to remove the human-made carbon dioxide in the atmosphere. Time is of the essence.

Technology has changed our relationship with time; our sense of time shrinks with each new invention. We not only live in the Age of Information but also the Age of Speed. How has your personal sense of time changed over your lifetime? It is apparent in how agitated we become when the Internet connection is too "slow" or when web pages do not come up at lightning speed. We check our e-mail, surf the web, and send multiple text messages to friends at speeds faster than most of us can react to what appears on the screen. Given this propensity for speed, it will be very difficult to come to terms with any problem like greenhouse warming, which will be with us for decades, if not millennia.

Our technological prowess has led to situations in which problems build over time until they reach a point of catastrophic consequences. Toxic pollutants accumulate for years before they have a significant effect on health. Oil spills go unnoticed, or their effects remain hidden in the oceans for years. We are not well designed to deal with problems that have these built-in time delays; it is not surprising that grappling with these issues is challenging.

These pressing issues force us to redefine our relationship with time. We need to stay mindful of the present but at the same time see the implications of our current actions for the future. If we were truly attuned to the present we would implicitly care for the future. We would be conscious of what we were doing and what the impact of our choices is on future generations. As such we need to embody a new sense of time to address critical issues that extend out into the future.

How do we relate to the physical nature of our body and those of others? Given our level of materialism, one could argue that we are very sense oriented; our consumption of things must mean we are very aware of the world of matter. Our reckless consumer behaviors belie this argument: we have a very poor relation to material objects given that much of what we purchase ends up in landfills after a short time. I believe we have lost our ability to relate to matter in a caring and concerned way. The explosions in obesity and type 2 diabetes are indicators of the lack of care with which we regard our bodies. The prevalence of body dysmorphia is yet another indicator of this.

Throughout much of Western history, one dominant philo-sophico-religious theme has been that the body and its sensual pleasures pull us away from God. Sensual pleasures inevitably lead to sex, and in a male-dominated society, this also creates the "problem" of the female. The patriarchal West has never really figured out how to hold the feminine in a positive, cre-ative manner. Our cultural complex presents the feminine as temptress and as being less than the masculine. This view, in-culcated in us from the story of Genesis onward, is alive and well in our current social system. News stories abound with the mistreatment of women around the world; in some places bodily disfigurement is the accepted punishment for perceived transgressions. If we are to move forward in creating a flour-ishing future, we will need consciously to hold all dimensions of the feminine and masculine. All of these dimensions are needed to understand being in the world.

Much of our "being in our body" is unconscious. Our hearts beat and lungs breathe in and out without our conscious awareness. Our bodies withdraw from pain without conscious reflection. Our hunger for food and sex is followed, not pre-ceded, by conscious awareness. We often ascribe such desires to our animal nature, suggesting that our instinctual body is lower down on some metaphorical chain of being. I believe this pejorative perspective of our inner instincts extends out to animals in general and is reflected in our callous or uncar-ing destruction of many species in the world.

Our physicality builds bridges beyond our bodies. The more we attune ourselves to this reality, the more we are in tune with the world around us. If we immerse ourselves in a state of sensed connection, then acting destructively toward the world becomes more difficult. We would feel so much a part of everything that in doing harm to the world we do harm to ourselves. This sacred state of sense creates an empathic embodiment, placing us in balance with the world. In denying ourselves this level of connectedness we cut ourselves off from the world, and the result of such disconnection leads to environmental crises.

The three qualities of being in the world, space, time, and body, come together in the fourth dimension of relationship, which places us in the world as an engaged participant. We are always in relationship, which includes our inner as well as our outer world. In any relationship there is an exchange between I and other. I stand before a tree, and the tree stands before me. I observe not only the form of the tree's bark but hear the wind blowing through its limbs, smell the aroma of the bark, feel the texture of its surface. When I place my hand on the bark of the tree, this living being reacts to my presence. Relationship is rooted in this back and forth between two beings.

What of relating to another, human being to human being? I sit across a table from a friend with whom I have agreed to meet for lunch. This decision, by itself, provides a definite space for our being together. Our relating occurs in the presence of a shared meal, a setting very different from sitting in a park or bar. The mood of our surroundings affects our being with the other. The presence of the wood table, the food on the table, and the sounds around us all create an atmosphere that influences the quality of our relating. There is a resonance between mood and being in the world that arises from this particular environment, opening us to the realization that our relating to each other is not separate from the world surrounding us. We realize that as our environment changes, our relating to the other changes. We attune ourselves to the other, a process both conscious and unconscious. Sitting with my friend, I focus on what he is saying, and my reactions to his words, facial expressions, and tone of voice are often unconscious. His words may make me feel happy, sad, concerned, or even unconcerned. My mood is modulated by his mood, whether I want it to be or not.

Bringing consciousness to the moment, I begin to sense how my emotions react to his world. I shift more to the center of my being in relationship with him. My experience of relating to him is important; I learn that by attuning myself to him life is enriched.

The mood with which we engage in a situation can vary greatly. We may feel joyful and fully alive to what is going on, or we may feel morose and removed. Mood affects our being in the world. The mood of the environment influences us in unforeseen ways: we enter a room, and our mood changes based on what is taking place in the room. Imagine being fully engaged in a situation, perhaps with someone we find attractive, or participating in an event that requires our full attention. We are strongly connected to the people, the action, and the physical environment surrounding us. In such a situation we are in a state of attentive care toward others. This type of relationship sustains itself for a time because all of the participants are engaged in the moment. As the surrounding environment changes, however, this attuned moment may fade away, leaving us with a yearning for that lost experience.

This description of our relationship to the world through space, time, and body show how our surroundings affect us and how we affect them. Descartes' dualism dies when we awaken to how immersed in our worlds we actually are. At this point, we can no longer sever ourselves from the world picture. We must admit that our being and the world's being are entwined; our lives become richer when we awaken to this fact. I believe we must remember this fact and manifest it in our actions.

A sense of mystery accompanies this type of relationship, in which we are never sure what will be unveiled in a given moment. When acting creatively, we don't know where things will end up; often what is happening is beyond words. The poet John Keats described this as "negative capability": "when [one] is capable of being in uncertainties, Mysteries, doubts, without any irritable reaching after fact & reason." In this type of relationship we are able to tolerate and revel in a sense of not knowing. We are attuned to the situation and able to al-low things to unfold in whatever manner they wish. We invite mystery into our experiences. Deeply affective relating is often full of pregnant pauses; it is a holding environment that waits without anxiety. In such relating, trust becomes integral to our process. True craftsmen trust in their skillfulness and wait for the moment to work co-creatively with what lies before them. They wait in openness to bring forth the new.

Of equal importance to how we live in relationship to the outer world is the relationship with oneself, for relationship exists in both outer and inner worlds of being. Without a car-ing, creative relationship with ourselves, relating positively with the outer world becomes a challenge. Turning our gaze away from our inner selves often allows us to project our shadows onto the world, distorting our relationships. We see through a glass darkly. To address this possibility, it is impor-tant that we look within. How does our level of self-care affect our ability to heal the world? Are our valiant attempts to save the outer world related to our inability to tend to our inner lives? These are important questions to reflect upon in our at-tempt to avoid destruction in the world.

How are these thoughts on relationship relevant to the issue of global warming? If our relationship is rooted in a sense of attunement and caring, then we truly value the material world around us. We value the lives of others who are affected by our actions. Our sense of relating would be empathically in tune with the world, which would mean that our decisions would include not just our interests but the interests of others. We would be unable to mine the tar sands of Canada because we would know that this decision brings destruction to many. We would seek solutions to our energy needs that would be creative and fulfilling for many rather than for the few. We would build a world with the same love a craftsman employs in creating fine work.

In essence, I am speaking of living authentically in the world. Authenticity does not come easy in a world filled with so many distractions. Technology draws us into its world of collective approval seeking. Through the power of image and text messaging we find ourselves wanting to be a part of the whole. Our inner sense of self, who we are as being in the world, becomes entangled in the Internet's definitions of who and how we should be. When we cross that fine line of paying more attention to the not-so-subtle messages of our cell phones, iPads, and laptops than to the voice of wisdom within, we lose our authenticity. Living the authentic life unifies our sense of experienced time, space, body, and relationship. In the act of gathering our being in the world, we live in harmony with ourselves and with everything around us. The more authentic we become, the easier it is to live in balance. During this time of increasing disruption from global warming each of us is called to become more authentic in our being. If we do not walk a path of authenticity, we will lose much of the world we have known for millennia. If we do walk the path of authenticity, we open ourselves to inner and outer beauty.

Beauty's Way in the World

When we awaken to the call of Beauty, we become aware of
new ways of being in the world.

—John O'Donohue

As I walk along a path to a waterfall in Big Sur, a sense of expectancy arises in me. Pausing, I look down over my left shoulder at a pristine beach, a turquoise ocean, and a stream of glistening white water falling from the rocky heights above, a vista from Paradise. People with cameras attempt to capture the beauty of this place, which has placed us all in a mood of reverie and joy. How is it that the beauty of place plays such a significant role in determining mood? I continue to walk the path, looking out onto more wondrous scenery, the falls, the vast, open ocean, the expansive blue sky, and the large palm trees rising from the mountainside. Beauty lies in every direction. The being of this place seizes me. Nature unfolds its unique beauty, and in its presence I am transformed.

Such offerings of beauty are not limited to the natural world. Upon entering Grace Cathedral in San Francisco, the first thing I notice is the silence, which is unlike that of a library, hospital, or even a small chapel. This is an expansive silence that pulls me into its vaulted, voidlike, sacred space. Glowing, diffuse, multihued light, originating from stained-glass windows high above, flows into this immense interior. In a state of obeisance, my eyes follow the soft light to the floor; embedded in stone rests an engraved, circumambulatory spiral labyrinth. In this moment I am immersed in the numinous nature of this space. Clearly, beauty resides here in this human-hewn creation.

Beauty dwells in the realm of the senses, opening us to numinous experiences. It transforms us and elevates our spirits. To experience beauty fully we must gaze deeply into the world and look beneath the surface. Gazing into the interiority of the world brings a sense of aesthetic wonderment and interconnectedness. In beauty we see beyond the visible into the invisible nature of the world and connect to the things that most matter to us. I call this seeing deeply.

We all have the potential for seeing our world this way, but we often forget that we do. We can, however, reawaken to this deeper way of connecting with our world. To accomplish this, we must remove the filters clouding our awareness. In this process, we can see how our psychological and social environments refract our experience of the world. What we learn from parents, teachers, and peers affects our ability to see the world anew. These social norms and belief systems create filters for perceiving and making sense of the worlds we inhabit. They help us live safe and organized lives, but this comes with a price. Norms skew our perception of the world: they place another's understanding between us and the world's beauty.

There are many ways to experience beauty in the world. Our senses provide unique and nuanced approaches to beauty, but our senses are not the only ways to know the world. Science provides a way to see beauty in the world as well. The elegance of the theory of gravitation, how theory unifies observations of the world, or the mathematical succinctness in explaining the evolving universe are all forms of beauty. For centuries we have looked at the world, seeking to understand it. Our curiosity about the ways of the world begins with the senses, but we seem to yearn for more than a collection of observations. We want to "make sense" of the facts; we seek to order and understand.

Using our intellect, we unveil the multiple facets of the natural world, and through this process we perceive order and predictability. This process of "looking" at nature also opens us to its beauty. The fact that we can write mathematical expressions that provide an accurate description of nature is astounding. Newton gazed upon nature and arrived at the universal law of gravitation; applying his mathematical understanding to the solar system, he was able to derive the motion of planets about the Sun. A single, compact, elegant mathematical expression unveiled nature's means of keeping the heavenly bodies eternally bound to one another. My love for science rests in a feeling of appreciation for the beauty of how our mind sees order in the cosmos.

My colleagues and I are gazing at a computer screen, across which are flowing multiple colorful threads. The filaments of each color form tight patterns, like one sees in a stream. I am amazed at the beautiful dance of many-colored threads and equally amazed that this moving image has been produced by a computer model of Earth's atmosphere. By solving equations heirs to the ones created by Isaac Newton and other scientists of the past four hundred years, we are able to produce a replica of how the atmosphere circles the planet. The wavelike flow meanders around the globe. We see how these flowing wind patterns carry vapor from the warm tropics to the colder regions of the poles. My colleagues and I share in a feeling of accomplishment: this was a climate model built from our hard labor. I am moved by the fact that the human mind can employ the language of physics and the ingenious technology of computers to model the Earth's atmosphere realistically. Science is a beautiful thing.

Of course, the shadow side of science is its propensity to view the world in a purely dualistic, mechanistic framework, and this contributes to the view that if we can understand and predict nature, we can control it. Science as a means to control carries power, and the presence of power always places us in a moral dilemma. How do we use this power? Who controls it? Great temptation arises with power. Beauty retreats in the presence of power. The challenge for science is how to hold on to its unique way of opening us to beauty without sacrificing beauty to the seduction of power.

Why is seeing deeply important for our transformation? If we continue to see through the old story of separateness, then our thoughts and actions do not recognize the inherent value and beauty in the world. To create a flourishing world that respects life on this planet, we need to see the value of the material world surrounding us. Our actions must arise from this sense of valuing, not from disconnection. If we open up to the beauty around us, we are more likely to act from a place of *felt valuing*.

Seeing deeply attunes us to our world. It involves more than just accounting for what is recognized. I watch a couple walking together hand in hand, how they carry themselves, the way they look at the other, all of their gestures revealing deep commitment. In watching the couple, I become attuned to them and am pulled into their world.

This level of connectivity extends beyond a relationship between two people. Consider craftsmen working carefully to build a piece of furniture: they hold compassion as an integral aspect of being attentive and attuned to their work. In their choice of wood, the care for their tools, and the tidiness of their workshop, they stay attuned to creating beautiful, functional furniture. When this furniture leaves the workshop and moves to someone's home, the mood of the craftsman travels with it. The mood of the craftsman dwells in the work, and the work unveils the hidden care of the craftsman. Beauty becomes enfolded in this process, providing us with more meaning.

Beauty and imagination open doorways to possibilities. Imagine a world where the quality and value of life is an integral part of how we measure a successful economy. One country manifesting this vision is Bhutan, which uses a measure of gross national happiness (GNH) to plan its economic future. Unlike gross domestic product (GDP), GNH accounts for quality of life and well-being. Imagine if more countries adopted this measure for economic planning. We cannot create such a world without first imagining it; a failure to imagine our future forces us to repeat the past.

As I walk that path in Big Sur, I think of the many people traveling it. They are here with me to experience this place of beauty, the rich flora's brilliant colors. The air feels warm, and I am attuned to earth, sun, and the sound of the waterfall ahead of me. The beauty of this moment makes me whole. I am conscious of how mind and body allow me this experience: my mind intently informing me of what I am seeing and leading me along the path, my body opening me to the sounds, sights, and tactile sensations of the surroundings. I experience this world as valuable to all sentient beings, human and nonhuman, like the lizard sunning on the wooden rail nearby. My imagination reaches out to others here, informing me of how they are feeling in this special place.

I am reminded of the Navaho blessing way:

I walk with beauty before me. I walk with beauty behind me.
I walk with beauty below me. I walk with beauty above me.
I walk with beauty around me. My words will be beautiful.

We need to weave this blessing way into a world of flourishing sustainability. When we live the blessings of beauty, we see deeply into our world. I recognize why this path needs to remain whole and present for all. I recognize that my actions must ensure that this path and place be preserved for generations to come.

Why Meaning Is Important to Being in the World

Meaning makes a great many things endurable—
perhaps everything.

—C. G. Jung

I enjoy walking along the beach. The boundary between earth and water puts me in a meditative mood and opens me to wonder about what it all *means*. Today, looking for life at the sea's surface, I spot an otter swimming in his watery world. I find true gracefulness unfold from his wily flip as he dives to unseen depths in search of sustenance. Returning to the surface clasping something tightly to his belly, with what seems to me nonchalance he floats on his back, forepaw rapping on a prized shellfish. A trickster seagull floats nearby, ready to seize upon any slip on the part of the otter. As the otter plays with his food, he keeps a watchful eye on the hovering gull. I laugh out loud. There is play here. Is the otter teasing the bird? Is the bird playing with the otter? Are these creatures performing for me in some way? In this moment, I am a part of their world and they mine. In some wondrous way I am gathered to their world. This gathering of my world and theirs brings some sense of meaning to my life; I am a participant in the natural dance of life. I value this moment.

Meditating on meaning opens us to wondrous worlds. We are beings who seek meaning in life. The Jungian analyst James Hollis writes, "We are the creatures who demand meaning, suffer its absence, and twist ourselves in painfully tropic ways to find its light." Even those who adopt a position that life has no meaning provide themselves with a "meaningful" answer to the question of meaning. It is impossible to imagine living in this world without a sense of purpose, goal, direction, or reason for being here. The importance of meaning is evident in the fact that whenever we lose a sense of meaning we feel alone, lifeless, and depressed. Meaning stabilizes our lives and grants us an opportunity to express ourselves creatively. Near the end of his life, Jung emphatically stated that we "cannot stand a meaningless life."

Usually, meaning involves making sense of the world. It may even inform us of our place in the greater cosmos by defining our relative relationship to the world at large. Meaningful symbols create containment or a holding space for us in our lived world. For many, a religious symbol provides a sense of order and peace; others may find meaningful symbols in the scientific explanations of our origins and function in the world. Those who find meaning in their lives know why they are here and have a sense of purpose and intention. Without a sense of meaning, we fall into a feeling of isolation and separateness. We lack a sense of belonging.

Meaning arises from a multitude of sources. When we are young, our parents shape our sense of meaning, at least for a while. By the time we reach adolescence, we may begin to reject these parental meanings and venture out into the world to find what makes sense to us. During this time of life, we rely on either peers or institutions for guidance in defining our meaning in the world. Traditional religions—as well as nationalism and militarism—may also provide meaning for us. These externally imposed sources of meaning provide us with a sense of place in the world and perhaps a strong feeling of community and purpose; however, they often begin to lose luster and stop providing us with a sense of intention or direction. When we lose our sense of meaning, we experience a malaise or become melancholy about the purposelessness of life. At such times, turning within to find meaning may occur, and, seen in this way, our state of meaninglessness actually can awaken us to significant transformation.

In Jungian psychology, meaning may arise through identification with entities from the outer or inner world. Identification with something or someone implies that we have allowed the "other" to define us. Identification is an unconscious forfeiting of who we are meant to be. In certain moments of clarity, often in the midst of discomfort or suffering, we become aware that something is not "right" with life, and we begin to realize that the meaning of life has been lost or borrowed from another. We realize we are living an inauthentic life. We begin to question our meaning. Such moments are invitations to finding deeper personal meaning to our lives. Such discovery does not come easy and involves throwing off our identifications so we can consciously recognize who we are.

As discussed previously, cultural complexes arise from our values and beliefs. These grander scripts—often most evident in the form of laws or norms of social behavior—provide meaning to societies and nations. Meaning founded on cultural complexes helps provide cohesion and stability. As religions they may put us in touch with the transpersonal numinous. My feeling of awe as I walked into Grace Cathedral was, no doubt, rooted in my childhood experiences in church, but it also touched something much deeper within me.

Our problem with overidentifying with cultural complexes is that we transfer so much of ourselves to a particular idea or belief that we are no longer able to make balanced, informed decisions. The shadow side of these cultural complexes is that our overidentification with them blinds us to reality. Identification is an abdication of individuality to something other than our selves; we sacrifice the ability to see the world from our own particular perspective. We lose clarity and see everything through the lens of the complex. We cannot make up our own mind about things but instead fall lockstep into the complex's view of the world. History is a running narrative of how mass identification with cultural complexes leads inevitably to destructiveness.

Archetypal patterns are at the root of many conditioned and cultural complexes. In ancient times they would have been the "gods." Archetypes are the universal ways that we see the world, and they manifest in images and symbols. When we see large groups reacting to a situation in the same way, with the same intense emotional charge, most likely we are witnessing the appearance of an archetypal pattern. These strongly affective patterns seize us and fill us with a tremendous numinous experience. Picture the enthusiastic response at a large rock concert or sports event, in which the crowd is catalyzed into cathartic release. Being touched by an archetypal pattern can be life changing. The music of the 1960s served as an initiation to a new lifestyle for many young people at the time. An image associated with such an experience can direct us in life. For many Christians, the symbol of the cross brings deep meaning to life; for others nature may do the same thing.

Archetypal patterns, as with complexes, may appear in the form of light or shadow. In fact, all archetypes contain opposite forms of expression. We would love to have these patterns of perception manifest in their positive forms at all times, but this is not possible. The darker side of the archetype is always present, and such negative patterns move people to dark behaviors. Meaning rooted in such shadowy patterns leads to destruction rather than creation. Major conflicts throughout history, from the burning of witches in the Middle Ages to the Holocaust in the twentieth century, are manifestations of collective darkness holding unimaginable power. Meaning derived through identification with such darkness is seductive and destructive.

Also, we derive meaning through identification with our status or position in life. All of us develop personalities that we present to our worlds. Our behavior and mood change, depending on whether we are with family, friends, or professional colleagues. The roles we play are dependent on what we perceive is expected of us, which is especially true of work situations. Our behavior generally changes when the boss enters the room. If we overidentify with our roles, or personas, then we allow the role to define our life's meaning. This is a dangerous situation to be in given that outer circumstances are always in constant flux. A new boss shows up, and we discover we must change our mask to work with him. Situations such as this are invitations to discovering our inauthenticity.

Our tendency toward overidentification is not limited to people and their relative positions in life; it occurs with material things as well. When our sense of identity—another form of meaning—depends on what and how much we own, we have given an object tremendous power over us. Advertising agencies are very aware of our susceptibility to granting material objects the power to define who we are, and we are in a precarious situation if our meaning in life is determined by what we possess. This is an unstable lifestyle because the material objects before us are forever changing—again something ad agencies use quite effectively. If we allow material things to determine our sense of meaning, we fall into a state of perpetual consumption, continually needing new things to feel meaningful.

Of course, some objects in the world need not distract us from finding meaning in life and may actually lead us to a deeper sense of meaningfulness. Art opens us to worlds of experience that may connect us to the numinous nature of life. Our sources of meaning are mirrored in the development of art throughout history. Consider the artistic creations arising in the Middle Ages from the Christian story, or the arresting landscape paintings populating the age of romanticism. Nature is another "thing" in the world that has had a profound effect on us, and many people derive great stability and purpose from their connection to the natural world. In my clinical practice, I have discovered that a disruption of this connection often leads to a paralysis in everyday life. In such cases, the path to healing often requires the wounded person to reestablish a felt connection to nature.

Ultimately, each of us is called to discover a deep, sustaining meaning in life that does not arise from unconscious identification but from a conscious commitment to finding the source within. I believe the most profound source of meaning comes when we touch the archetypal center of wholeness that rests within each of us. Jung called this archetype of wholeness the Self, which is quite different from the ego-self, or "I." This Self, as center of wholeness, encompasses our entire conscious and unconscious being; when we are in touch with this archetype, life feels fulfilled. When we are connected to wholeness, we find meaning in everyday life; we value others, ourselves, and the surrounding world. Our ultimate task in life is to discover this ever-present center and to stay in direct relationship to it. Lacking a connection to wholeness, we feel empty and search for meaning in the ephemeral world; we don't see the inherent value in the things of the world and fall into separateness. It is important to reiterate that what I am describing is a *conscious* connection to our inner world, not *unconscious* identification. When we are consciously aware of our inner world, we work with that world to live life more fully. When we are unconscious of our inner world, it works through us in self-defeating ways.

So, what does meaning have to do with our current environmental situation? What can we learn from questioning the meaning of the role we play in climate change? What role has identification played in affecting our situation?

Establishing our meaning through materialism and consumption is self-destructive. Global warming exists because of our excessive consumption of energy. On a per-person basis, the average American consumes more than twice the energy of a European. Energy consumption by the world is on a steep upward turn. Often global energy consumption is used as an important measure of human "progress." To address the problem of climate change will require us to redefine our measure of meaning beyond this single factor. As long as we view consumption as a significant source of meaning, we commit ourselves to a life of depletion and destruction.

Studies show that defining meaning through materialism leads to serious deleterious effects, including an increased sense of insecurity, decreased happiness, degradation of personal relationships, and poorer health. Materialistic values also lead one away from a sense of authenticity and autonomy. Given this vast array of the negative side effects of our current materialistic sense of meaning, why do we persist in this living myth? Unfortunately, we are a species known to persist in self-defeating behaviors. All of us have habitual patterns that we know are not contributing to our health and well-being. It is easy to point to another who is living a self-defeating lifestyle and analyze his or her poor choices; turning this analytical gaze on our selves is far more difficult.

As individuals, if we believe that material possessions provide security or make us feel better about ourselves, we will stray far from finding true happiness and an inner sense of sustained meaning. Identifying with the material "thing out there" to define who we are is fated to failure; what is out there will forever be changing. As long as we misidentify monetary wealth with true wealth we will be caught on a treadmill of accumulating more and more for less and less, repeating the mistakes of Earth-Destroyer in the ancient Greek myth. Psychologically, we are caught in the compulsive repetition of an archetypal pattern. Our current living myth of unconstrained economic growth pushes the world ever closer to catastrophic collapse. Eternal growth must give way to meaningful sustainability.

We are called to make sense of our being in the world; we are compelled to find meaning in our seemingly limited situatedness. Historically, we have met this call by defining ourselves in terms of something greater, perhaps reflecting our innate need to be contained or held by something larger than ourselves. Our development as individuals and as societies begins with the search for a relative relationship between the small and great, the limited and infinite, the microcosm and macrocosm.

At the beginning of civilization, we found the greater outside of ourselves in the form of gods. In the Western world of the seventeenth century, we began to replace these outer gods with an inner sense of supreme subjective reason. Today, we live in a world where the greater is defined by economic growth and the accumulation of wealth and power. In our process of "development" we moved farther from a sense of meaning that allowed us to feel like an integral part of the greater. Our current sense of meaning and its associated values lack the ability to fulfill us. We are living a myth that creates inner emptiness. The material things in the world, seen as dead and disconnected objects, will never provide us with a meaningful sense of the greater. Materialism fails to be a living symbol.

Overcoming this self-defeating and dying myth requires us to move beyond a meaning rooted in senseless materialism to one truly representing the greater, one transcending the "I" and even the "we." We must discover and connect with the transpersonal within ourselves, to rediscover our place in a living, dynamic cosmos. Here the transpersonal is not just spiritual but also the greater social good, the animate natural world, and the sense of holistic well-being. We are tasked with finding our own sense of the greater within; only then can we see that our attempts to fulfill ourselves from without are doomed to failure.

This goal cannot be achieved by repeating the past. We have always possessed objects, and they have served us in various ways. What has happened over time is that the possessed objects have come to possess us. We cannot look for the transpersonal in an outer form, whether a material object or a concretized image of an external god.

Watching that playful otter and seagull connected their world to mine. The vastness of ocean and sky placed me in the presence of the greater, and in that moment a door opened to a meaningful experience of being in the world. In such times, what we own, and how much we own, no longer define our meaning in the world, and we become bigger; our lives become fuller. The desire to possess and the possibility of being possessed fade away.

IV

Awakening

How Our Many Worlds Are Entwined

I maintain that it is no more possible to know the parts
without knowing the whole than to know the whole without
knowing the parts individually.

—Pascal

In Point Lobos Reserve, I spy a slender woman peering through
binoculars at a tree. Seeing me, she raises her hand silently
and points: the tree is full of vultures—red heads and large,
dark, hunched bodies, arranged as if they had drawn straws
to see who would be at the top. The birder, for surely this is
who she is, notes that the vultures are no doubt here because
of the harbor seals. I walk on a bit and gaze down at a beach,
where a group of seals stretch lazily out on the sand; some are
completely black, others speckled. Occasionally, one or two
of them swim out into the secluded waters of the cove and
then return to the beach, playfully going to and from the water. Suddenly, a vulture takes flight, and all twenty or so seals
undulate swiftly to the sea. In an instant all I see are a random
scattering of seal heads poking out of the water's surface, their
whiskered and alert faces looking like a pack of dogs.

What happened? Was it the aerial motion of the vultures,
those feathered forms of finality, that forced the seals to flee
to the sea's safety? I realized how interconnected life is in this

small inlet, how tree, bird, seal, sand, and sea support a vast ecology of life forms. If one being moves, all others react in an intricate dance. Before me is the entwining of living worlds. The interconnections between nature and our human-designed world are manifold. Ecosystems comprise a rich hierarchy of interactions that enable many forms of life to coexist in dynamic balance. At each level of a complex natural system reside smaller, intricate subsystems allowing the flow of energy and nutrients to sustain these entwined microcosms. Energy flows into the system and is distributed in optimal ways to ensure a stable environment. There are intricate communications and interactions among all who inhabit any environment. From bacteria beneath the ground to the leafed branches of the trees reaching into the sky, everything is connected.

How does the system "know" how to build these intricate interactions? Science tells us that coherence emerges out of such complexity. The system itself behaves in such a way as to give birth to structures that live in and organize the greater whole for the purpose of sustaining the environment for all. The interconnectedness of individuals becomes the means by which the greater system sustains and perpetuates itself. Such systems react constantly to changes in the flow of energy entering the ecosystem and to variations taking place within it. Changes in rainfall affect life in a forest, and any change in rainfall filters through the whole system. If the change is not too severe, the system will adjust. If the change is too large or long lasting, the system may collapse. The resiliency of the whole works only so far in its ability to preserve and protect the vitality of life. Interconnections do not guarantee a secure future, but they do imply a cascade of changes throughout the whole. Interconnection assumes that each part, no matter how seemingly small, plays a role in the governance of the whole.

This same dynamic extends to human-constructed social systems. These also contain systems within systems, in which one may fit within an even larger one. Consider the hierarchy of neighborhood, suburb, city, county, state, and nations. Socially bounded constructs including families, socioeconomic classes, religious beliefs, and political affiliation exist within each of these political and physically bounded systems. Just as with natural systems, energy and information flow through all of them. Of course, there are important differences between human and nonhuman systems, perhaps the most important being the level of conscious engagement. We have the unique ability to transform radically our environment at a level unimagined in the nonhuman world. While there are certainly many instances in which animals manipulate their environments, these pale in comparison to what humans are capable of. Perhaps the greatest change over human history is the magnitude and speed with which we have changed our global environment. So, unlike the natural world, which reacts mainly to external disruptions, we ourselves are the source of the disruptions to our world.

How do these reflections on connectedness relate to a transformation of consciousness? Consider that collective consciousness results from a flow of energy and information within our social systems. As it is a complex system, coherent structures arise to hold us together and guide us into the future. These meaning-providing structures may be beliefs, ideas, or ideologies. Currently we are in a situation in which structures created in the past have become counterproductive to our future survival. These old constructs provide support for a few, but in the long run they are severely limiting. Rigidity opposes resilience, and these old constructs are creating a system susceptible to collapse.

Recognizing interconnectedness is extremely important in dealing with the problems in our world. For example, the U.S. populace is periodically polled on what the issues of highest priority should be for the government. Improving the economy is traditionally recognized as highest, followed by lowering unemployment, lowering the deficit, and fighting terrorism. Addressing global warming is often at the bottom of the list. These results are in stark contrast to other polls that show an overwhelming feeling among the public that protecting the environment is very important. This apparent contradiction exists because issues closer to the day-to-day demands of life are deemed more important than the environment.

The American psychologist Abraham Maslow's "hierarchy of needs" model may help explain these disparate findings. After studying individual and social needs among Americans, Maslow proposed that our perception of needs fall along a vertical hierarchy, with the most basic needs located at the base, followed by needs related to family and community, with transpersonal experiences all the way at the top of the hierarchy. Maslow pictured this hierarchy as a pyramid of perceived needs. He argued that until our more basic individual needs like food, shelter, sex, and safety are met, we invest less attention in the higher needs. If we feel that our basic individual needs are threatened, we will focus much of our attention on them. The contrast in the polls between these priorities and protecting the environment indicates that people are concerned about their basic individual needs. They may feel that the environment is important in a more abstract, collective sense—as in *we* should protect the environment—but are far more affected by their individual fears around the loss of basic needs. Further research on Maslow's hierarchy has found that different cultures have differing orderings of needs. In particular, Asian cultures place the needs of the community at a more basic level than individual needs.

How does Maslow's hierarchy of needs relate to the interconnectedness of the world and the challenges of addressing climate change? We may choose to look at improving the economy, lowering unemployment, deficit reduction, improving national security, providing affordable energy, and addressing global warming as separate issues, but in reality they are fundamentally interrelated. Any comprehensive, enduring approach to addressing these issues requires a holistic view.

In a truly interconnected system, if you start at any particular nexus point and follow the threads connected to it, you eventually reach every point in the web. As the naturalist John Muir put it, "When we try to pick out anything by itself, we find it hitched to everything else in the Universe." In thinking within a context of interconnectedness we unveil an ecology of global values.

Let us begin with energy, an important issue for most of us. By relying so heavily on fossil fuels as our source of energy, we have built rigidity—a constraint—into our social and economic fabric. The political and economic liabilities that come with using fossil fuels have forced the United States into international conflicts costly to human life and the national economy. The fluctuation and instability in the availability of oil causes ripples in the global and national economies; thus we can see how energy cannot be viewed as separate from security, the economy, unemployment, and the national debt. Consider also that our reliance on fossil fuels results in environmental disruptions that have tremendous economic and social costs. A single severe storm, like Hurricane Sandy, causes billions of dollars of damage, and in our warming world, the likelihood of these events will increase. All of these issues are interconnected; we can no longer view our various worlds as separate. In today's world of intricate globalization, nothing is unrelated. Knowing this to be true, we need to find solutions acknowledging these entwined worlds and build more resiliency into the global system.

Ironically, a warming world is a result of seeing the world through a lens of disconnectedness. Consider the effects of the hurricanes Katrina and Sandy, which crippled cities, destroyed lives, and left millions without energy or fresh water. A more resilient system would respond to these storms with greater adaptability. Resilient complex systems with greater interconnectivity have the ability to reorganize rapidly on multiple scales. Of course, we can build more resilient and adaptive systems with more careful engineering and social planning, but this is an old approach, one that uses our intellect to address symptoms rather than causes. We need to move beyond this toward a transformation in how we see and experience our worlds.

Transformation must start somewhere. Does it begin with the individual, or do we try to change the whole all at once? We need to recognize that change is going to occur simultaneously on multiple levels throughout our social system. In the Buddhist tradition, Indra's net spreads across the universe, with highly reflective jewels placed at each intersection. Whatever is reflected at one point of the net is seen throughout the whole network. Imagine jewels reflecting jewels in an infinitely recursive dance of points and lights. I believe this is how transformation can take place now; each individual is a jewel in the global social network, reflecting ideas, feelings, beliefs, and actions. Change a few of these jewels, and the whole will reflect the change. This is a good way to view many of the social transformations that have occurred through time—

Gandhi's transformation of India, for example, or Martin Luther King Jr.'s participation in transforming racial inequality in the United States.

On a personal level, we must become more conscious of our actions and thoughts. If we continue to view ourselves as separate from others and the world, we continue to reflect an extremely limiting and narrow view.

How do we break through this old way of seeing the world? Do we have the capacity or potential to change our way of seeing? I believe we do: It is our inherent ability to tap into our tremendous capacity for empathy or compassion for one another and the world. Carl Jung states that what is needed more than ever in human history is to "temper our will with the spirit of love and wisdom." Unlike physical systems, where exchanges occur as flows of energy and mass, social systems include the flow of the critical element empathy. Our ability to care, have feelings, and value others radically changes the entire process within social systems. In connecting to our capacity to care, we can effect tremendous creative change in the world. This is the true key to our transformation. How do we connect to this basic ability to care? How do we reconnect to the spirit of love and wisdom?

The importance of empathy for others becomes most apparent when it is inhibited. In such cases, the social network suffers, and all who are in the web suffer. Empathy is a grand connector across our complex web of social interactions, from the family unit to the national unit. If we are to transform our global system and create a flourishing, healthy world, we need to make the flow of empathy a priority throughout our complex social systems. Currently, our emphasis is on the flow of currency and material goods, and while these are necessary for a viable and vibrant society, without the flow of empathy, our decisions do not account for the effects of our actions on the whole. The flow of empathy is the "glue" that holds the whole together. It is the property of life that creates our connectivity to the seemingly disparate parts making up our wondrous world.

Individuals are also a system, as I have noted throughout this work. Jungian psychology views the mind as a complex system including both conscious and unconscious processes. If transformation is to take place, does this mean that every individual on the entire planet must transform before we can address issues like global warming? Jung states, "The unconscious produces contents which are valid not only for the person concerned, but for others as well, in fact for a great many people and possibly for all."

We can understand these words in many ways. The universal patterns of perception, or archetypes, connect us through our collective reactions to images and metaphors. Certain images galvanize whole societies and lead to significant shifts in behavior. On the instinctive side, empathy acts to connect us as a whole. In the current world of global interconnectivity, there are also conscious interactions that facilitate unification. The implications of Jung's words are that global change can begin locally and propagate quite rapidly, a phenomenon that became popularized as crossing a "tipping point."

I imagine interconnectivity as a symphony orchestra, each instrument tuned to the others, their coordinated playing creating a patterned work of beauty. In following a score they are connected together in order to produce a melodically coherent creation pleasing to the senses. Often this cooperative creativity elevates the listener to a feeling of intense numinosity. If each musician is especially attentive to the pitch of his or her instrument and is mindful of his or her role in the symphony, then the interplay works in synchrony. The musicians flow together to create a whole that appears as a single, seamless entity. A resonance occurs in body, mind, and heart in a well-performed symphony, and with similar care and attentiveness this is what we can achieve in our social engagement with the environment.

I am particularly fond of the fugues of Bach. The fugue has an interesting structure: a single melodic theme is introduced, and then in counterpoint the theme is repeated in different tones by other parts of the orchestra. This contrapuntal echoing back and forth among the various instruments and musicians leads the listener into an expansive reverie. Fugues may contain a single theme and response, or they can be more complex, with multiple thematic elements and multitoned responses. They hypnotically circle around in subtly shaped form, musical mandalas leading one on a journey of beauty and, possibly, self-discovery.

I see our arguments on climate change as fuguelike exchanges of a single thematic element. We have experienced the opening theme, which states that we need to do something about global warming. Other groups have restated this opening statement in varying voices. The repetition of the original theme plays across a number of distinct social groups. Round and round the argument goes, with little promise of closure, for just like a fugue, the discussions on climate change never come to a close, continuing on with false endings that sound very similar to the beginning. We are caught in a fugue without a finish.

There are two key elements to any symphonic form. First, the musicians have a score before them, informing each what and when to play. The score is the grand map of the symphony, the creative fruit of the composer, and it provides coherence and form to the notes, chords, and tempo of the whole work. Without a score, the musicians would be present but lacking a *telos* or goal, lacking anything to lead them into the creative act of playing together.

Second, an orchestra has a conductor, who ensures that the musicians are coordinated in their creative process. The conductor interprets the score within certain constraints; for example, they define the tempo of various parts of the score and may emphasize particular aspects of the score.

Are there analogies with respect to these two factors regarding global warming? Currently, we have no score for addressing this issue; we lack a map that brings coordination and coherence to it. For decades the United Nations has attempted to write such a score; this approach struggles to find a score agreeable to all. The musicians want to create their own scores, leading to tremendous dissonance. Perhaps rather than a full symphony we need a small quintet composed of the top five emitters of carbon dioxide.

What would a score for such a flourishing future look like? Composed in the form of a fugue, it would contain an opening statement for a flourishing future for our children. In essence, we would have thought through and articulated the world we want to create for the future. This thematic statement would then be repeated by each aspect of society—in its own tonal form—amplifying in imaginative ways the creation of a future for generations to come. In the process of this contrapuntal exchange of imaginative development, the score would reveal the inherent interconnectedness of the world. As the skilled musicians in this symphony of creativity, we would address the issue of global warming in a coordinated fashion. Physical scientists, engineers, social scientists, business people, and those from the humanities and arts would create a holistic score for our flourishing future. We would be caring craftspeople of creativity, mindful of our place in the orchestra of life.

Imagining our entwined worlds as being akin to a great symphonic work informs us of what we need to do to move forward on the issue of global warming. Part of being a careful musician is to develop an ear for tones. What is the pitch, the tempo, the chord that we perceive? Similarly, we are asked to listen to what the other is saying about climate change. We are asked to attune ourselves so that we harmonize and resonate with others. Do they resonate with us? Is there dissonance in their speech? How do we bring harmony to discordant voices? These questions are important for improving our awareness of interconnectedness.

What are the key elements to working from a perspective of entwined worlds? First is the ability to take time to step away from the particulars and see the whole. We so often focus our attention on a narrow aspect of the whole and thereby miss the many connections existing in it. We need to develop our sense of reflection on the greater world.

Second, we need to develop our ability to hold the whole and not just the pieces. Often we feel we cannot grasp the varying aspects of the world's complex nature, yet to address our problems holistically we need the capacity to hold the center in the midst of this complexity.

Third, we need to experience a sense of the whole and its parts. We need to move away from looking at the world from a purely thinking mode to one of feeling and experiencing the world in a more holistic manner. Can we develop a sense of what people in the developing world are experiencing because of economic globalization? Can we begin to recognize our contributions to environmental problems in distant lands?

Fourth, we need to become attuned and attentive to the ways that our actions affect the world. We must awaken to how our local actions spread out to distant places around the planet. What we purchase, our energy consumption, and our daily choices do affect those around the world. The forces of globalization have insured this strong social connectivity. We must recognize our interdependence of being in the world.

In my local café, I look around and see people from all walks of life, some intently staring into their laptop screens, others in conversations about personal or business matters. The barista is busily preparing coffee, and the room is animated with energy; it is the beginning of another day for everyone. I gaze at my coffee cup and see on its curved white porcelain surface reflections from around the room. I reflect on the interdependencies embedded in this cup, on the people who made it, perhaps as far away as China. I imagine them going to a factory every day, no doubt thankful for being employed and able to support their families. Their act of creating this cup has reached me today in this café. I think of the clay used to make this cup, brought up from the earth and transported to the factory. This cup connects me to earth. The cup also includes the people who transported it to the store where it was purchased. It includes the fuel used in the various vehicles transporting it, fuel that also comes from the earth. I think of the people employed in diverse places to produce this fuel, which changes our atmosphere when burned.

This single cup enfolds and entwines so many worlds. Similarly, entwined worlds exist within everything, and in each moment of life we can choose to open our eyes to this. The more we open ourselves to seeing our interconnectedness the more we will feel connected to our world, and our actions will reflect this sense of connectedness.

Recognizing the Importance of the Transpersonal

A radical inner transformation and rise to a new level of
consciousness might be the only real hope we have in the
current global crisis brought on by the dominance of the
Western mechanistic paradigm.

—Stanislav Grof

In my early twenties, I spent time in the state of Washington
as an intern working at an experimental nuclear power facil-
ity. This was my first time in the West, and the flat, dry terrain
came as a great surprise. I was reminded of scenes from old
Western television shows and expected to see cowboys rid-
ing over the horizon. Amid this desert environment was the
small city of Richland, where I spent the summer. Often, on
weekends, I would relax in a public park that stretched along
the broad, lumbering Columbia River. One hot day, sitting beneath a large tree, I experienced a deep transcendent state of
being. In a moment, the "I" had melted away, leaving an experience of the most profound sublimity. There was no feeling of
separation from the world: everything was a part of me, or "I"
was a part of everything. This profound state of centeredness
seemed to last for a very long time. Eventually, I returned to
my old awareness, in which again I was separate from everything, yet the memory of that transformative moment stayed
with me. It was the opening of a door to a different reality, one

I have come to see as no less real than this world of apparent separateness. I now realize that transformation is at the foundation of our being in the world.

Transformation may involve changes within an individual, across an entire society, or, even, like my experience in the park, an engagement with the transpersonal. Often, the most profound transformations involve all three levels, from the individual to the transpersonal. I believe these transformations are more frequent than we realize, and I find a sense of hope in those I have experienced in my life. In terms of social transformation, I remember vividly three times in my life when people faced very challenging problems that were clearly recognized as morally wrong and appeared to be insurmountable. Yet each was surmounted.

The first time was the height of the Cold War, when I was in elementary school. I remember walking to school each morning not knowing whether this would be the day someone would "push the button," bringing an end to the world. Newspapers were full of stories about the possibility of nuclear war, and the overall feeling was one of tension, anxiety, and resignation. Many were resigned to the *fact* that the two superpowers would continue to build more and more weapons and that one day something bad would happen. In the midst of this anxiety, there were a few people who resisted resigning to this fearful fate. These individuals were from all walks of life, and they worked tirelessly for arms reductions. Interestingly, scientific research on the climatic consequences of a nuclear conflagration was a key motivation in the step back from the brink of destruction. I was one of the scientists who worked on the possibility of a "nuclear winter" resulting from a nuclear exchange. This was my first foray into an issue where human actions had global climate implications. The mere idea that a nuclear exchange could have consequences for the climate was very controversial at the time. I remember traveling to Washington, D.C., to participate in a meeting at the National Academy of Sciences on the issue. This was an eye-opening experience for a young scientist. I saw how critical scientists could be of one another and how important it was to be able to explain your factual evidence to others. Eventually, the superpowers woke up to the insanity of mutually assured destruction (MAD) and agreed to begin disassembling these weapons. Of course, the world still has nuclear arsenals, but the global threat of nuclear weapons is not as serious it was in the 1950s.

My second memory of transformation in the face of collective resignation was the reality of racial discrimination. In the late 1950s and early 1960s, there was rampant segregation throughout large regions of the United States. I believe that most people who looked around and saw what was going on knew this was morally wrong. Many people felt the problem could not be resolved, that it was too entrenched in the psyche of our nation, and that political resistance to change and deeply rooted racist behaviors were too powerful to overcome. Again, a relatively small group of individuals banded together to fight for civil rights. Dr. Martin Luther King Jr. and other brave individuals worked tirelessly to transform the nation. Has racial discrimination been eradicated from this country? No, but compared to the late 1950s and early 1960s, a radical transformation has occurred. In a matter of a decade, change became a reality, in spite of the feeling that transformation was impossible. Of course, the recognition of racial discrimination extends beyond the borders of the United States. South Africa's movement to eliminate apartheid is another example of how moral concern can lead to great social transformation.

The recent troubles experienced in U.S. cities reminds us that we must be forever vigilant with regards to racism.

My third memory is that of the Vietnam War. The memories of this event are palpable for me; I was of draft age during this conflict. I knew people who went to the war, some who did not return, and others who returned disturbed and confused. The suffering and anguish surrounding this so-called police action divided the nation. This was a time of profound polarization in the nation, rivaled only by what is taking place today. The nation, communities, families, and even individuals were split over the moral imperative of the war. Once again, grassroots activists took to the streets to call for transformation. The energy that flowed through people at this time was exhilarating, the single goal of ending the war unparalleled for the time. In the face of tremendous opposition by powerful political forces, people organized and fought for change. These three memories of mine reveal how transformation can occur on the social level through well-organized, focused dissent.

If we look further back in history, we see other times when great transformations took place in a short time. The independence movement in India began with a small group who recognized that change must occur in their country. In particular, Mahatma Gandhi led an entire nation of millions through massive social and political transformation in a relatively brief time. I'm sure that many people at that time said such change was impossible given the power of the British Empire, yet it happened. History is rich with such moments, all giving me hope that we will awaken to the moral imperative of addressing the threat of global warming and transforming our current world.

As in the past, once again we find ourselves in a situation calling out for transformation, a situation in which we know that continuing our uncontained consumption of Earth's natural resources is morally wrong. We know that the exploitation of the developing world to satisfy the perceived desires of the developed world is morally unjust. Equally unjust is burdening future generations with the immense disruption caused by global warming, given that they did not cause this problem. Indeed, we are in great need of transformation. As usual, the world is full of voices proclaiming the impossibility of such change because the fossil-fuel industry is too strong, politicians are too weak, and, worst of all, humans are too self-centered. If we choose to listen to these voices of negativity, transformation will not occur. Their voices are indicative of a strong, negative cultural complex that wants to preserve the old way of seeing the world. In terms of stories of old, these are the people who want to prop up the old ailing king. Fortunately, that complex always encounters resistance; there are always those willing to risk much to defy the rigidity of the status quo. As in the cases of nuclear armament, racial discrimination, and the Vietnam War, today we hear some arguing that rampant consumption and environmental destruction cannot continue.

This type of resistance to the status quo on energy production is useful because it keeps the issue alive and present; however, I feel that we must go far beyond these approaches to address the issue of global warming. An inner transformation must occur to create a flourishing world: inner transformation insures that any outer change will last and become the new dominant paradigm.

We require a transformation unique in human civilization, a transformation in which we relinquish our dependence on something to which we are highly addicted. It is for this reason that we need to look beyond a technological transformation and recognize the need for a transformation of our very being in the world. We need to look toward a transformation of consciousness; our behavior toward the world and toward one another needs to change. We need to recognize our interconnectedness.

We may view the emergence of the Internet as a technological sign of this yearning to be connected to one another. To date, connecting this way has been used for purposes both creative and destructive. Latent within the Internet is an ability to reach around the globe and communicate with one another. We are witness to its role in the recent revolutions in the Middle East, and the shadow side of this connection is in how religious ideologies use it to spread destruction. This is why we need psychological transformation to accompany our technological tools. We need more than just the technological ability to bring people face to face via computer screens. A transformation in caring about others has yet to occur on a global scale. We continue to see ourselves as sufficiently separate from one another, and this perpetuates our sense of emptiness. The only path forward that will fulfill us is caring for others.

Reflect on your feeling when you are with someone you care about and love. In these moments of close relationship we feel full and do not need something else to make us feel better. Our ability to relate is our greatest gift; now we are called to manifest this gift. We have reached a point in our evolution where we must choose between self-delusion and emptiness or self-fulfillment and creativity on a collective scale.

We are beings that create so much beauty in the world, and our capacity to love is boundless. Allowing this capacity to unveil and be expressed in the world is all that is required to create a flourishing world for our children. Our transformation of consciousness is tied to our ability to allow beauty to unfold within and outside of us. Our transformation of being in the world depends on allowing ourselves to experience our world in a richer sense of time, space, and body. This transformation is dependent on our embodiment of the creativity that dwells within us. The creation of the great cathedrals exemplifies our ability to bring beauty into the world. The ability of artists to evoke awe in us through beauty continually reveals our creativity. These structures were not built in a single lifetime but extended over many generations. This shows that we can look out into the future and stay connected.

Complete transformation occurs at the intersection of two ways of dwelling in the world; the most familiar, horizontal dwelling, is being a part of society. We are born into the world as social beings. Our families, friends, colleagues, and social groups surround us and help define who we are. Our social connectivity allows us to transcend our sense of individuality. This type of enthusiastic experience can be either positive or negative; we can get caught up in a collective excitement that allows us to feel our connectedness to others, or, from the shadow perspective, we can become an unquestioning follower and be inauthentic. In contrast, the vertical way of dwelling is one in which we connect to the transpersonal. Some may call the transpersonal God, others a Higher Power, Self, Buddha-nature, Tao, or Nature; there are many ways to describe that which is greater than the individual I. The transpersonal is more an experience or process than a thing. When Augustine was asked to explain time, he said, "I know what it is, yet when you ask me to explain it I know it not." I feel a similar answer can be given concerning any attempt to define the transpersonal. Anyone who has had an experience like the one I had when I sat by the bank of the Columbia River will know what I mean by the transpersonal, and if we fall into the trap of trying to catch and define it, in that moment we kill it. As in horizontal dwelling, with a connection to the transpersonal, an individual is merged within something greater than the individual. Horizontal and vertical ways of dwelling have existed for ages, and most of us have chosen to spend our time dwelling on the horizontal plane, immersed in our social worlds. In dwelling on this plane we dedicate ourselves to family, work, community, and nation. Others choose to spend much of their time dwelling on the vertical plane, dedicating their lives to spiritual pursuits. What we are called to now is to dwell at the intersection of these two planes, in which we transpose our orientation from the horizontal plane to the vertical without leaving the horizontal. The challenge of dwelling at the crossing of the two is to embody both states of being.

Our ability to follow the path of transformation depends on a connection to our inner archetypal world—the vertical—*and* the outer phenomenological world—the horizontal. Opening to the greater archetypal patterns in life, we find that the way forward is filled not with difficulties but opportunities. Jung felt that there was a central archetype of transformation, and this archetype often manifests at particular times, leading to the reorganization of an individual's life or an entire society. The ancient Greeks called such moments in time a *kairos*. Facing climate change as an opportunity to create a better world transforms our relationship to the issue. The appearances of spiritual traditions like Taoism, Buddhism, or Christianity are examples of such transformative moments. We are at a turning point in human civilization: an archetype of transformation is occurring; a new coherence is arising around our global interconnectedness. The transpersonal dimension is needed to support us in this process. Jung notes, "Everything now depends on man: immense power of destruction is given into his hands, and the question is whether he can resist the will to use it, and can temper his will with the spirit of love and wisdom. He will hardly be capable of doing so on his own unaided resources. He needs the help of an 'advocate' in heaven."

Who or what is this advocate? It is the archetype of transformation, the deeply rooted part of the psyche that works through us to accomplish a healing of both the inner and outer worlds. It lies at the intersection of social and transpersonal realities and enables us to create a new way of being in the world, leading to our flourishing future.

12

Awakening to One World

With our eyes open, we share the same world; with our eyes
shut, each of us enters his own world.

—Heraclitus

Walking in the Point Reyes National Seashore I come upon a
small cypress tree growing out of a bare rock outcrop. I mar-
vel at the tenacity of this lifeform, rising directly from such
a rugged environment. I consider how this tree has found a
sufficiently supportive place, allowing it to live in balance with
its surroundings. Seeing this tree's *being* in the world reminds
me how each of us holds the potential for growth and the possibility to achieve balance with our world. Can we take a lesson
from this solitary tree emerging from bare rock? No matter
where we are, no matter how seemingly hostile our environment, we can strive to manifest the deep value of life in this
world.

Considering these meditations on our changing world, it
is now time to weave them together to create a vision for our
future. We can think of this as a tapestry that unfolds a story
full of phenomenal patterns of change. The purpose of this
tapestry, like those of old, is to tell a story we desperately need
in these turbulent times, a story rooted in our ability to see the
inherent value of the world and directly experience its lumi-

nous beauty. It is a story of how we perceive the many worlds in which we dwell as truly One World.

What are these various threads composing our tapestry for the future? Interestingly, each thread mirrors an image of One World. Perhaps at the most fundamental level our worlds are united in matter itself. Everything we see in the universe emerged from the primal birth pangs of the Big Bang. Physicists tell us that just after the formation of our universe all matter occupied an infinitesimally small space, and at such distances each primal particle became interconnected, entangled with one another. Over the past fourteen billion years the universe unfolded into its present appearance, stretching to unthinkable distances, with recurring particle interactions creating new interconnectedness. The profound implication of this cosmic quantum reality—concerning all things material—is that everything in our universe, including us, is eternally entangled at the smallest of scales. What are the implications of this reality for the larger scales of humans, planets, solar systems, and beyond? This quintessential question may be the hidden element behind our new story. As has been noted by others, we are made of stardust, and at the most basic level of reality we remain connected to those stars. I believe human consciousness, with its phenomenal ability to apprehend the vast depths of the smallest and expanses of the greatest, is now poised to turn its gaze inward. In this process, we see deeply into the interior of all worlds and know that behind these many worlds remains One World.

At the biological level all life shares much of the same genetic material. We are truly united at this most fundamental life level. We carry the code of characteristics that allows us to know one another as a part of the grander panorama called

Life. Genes, however, are not the only way we are united. In the realm of the sociological we learn that our species has the unique capability to leap beyond our genes, communicating through language and symbols to construct coherent paradigms of collective behavior. Our beliefs, ideas, and cultural constructs unite us in ways that transcend all borders. These coherent patterns of values, beliefs, and stories may manifest as either conflict or cooperation. We are in a position to choose stories that focus our purpose on creation over destruction. The world is in need of more stories of creation and creativity. Cultural evolution allows us to act in a creatively caring way toward all life on Earth. Imagine seeing the inherent value of the world rather than its monetary worth. Imagine belief systems that elevate and unveil the beauty of life on Earth. If such luminous lenses are used to view all aspects of our world, each action will arise from an inherent sense of connectedness. Our decisions about energy, lifestyle, and consumption will be naturally enfolded into our being as individuals and as societies. Our *telos*, or purpose, would be to strive for the better of all, not just a single individual or particular group.

But what of the forces at play in the veiled background of our unconscious? What of the psychological threads used to weave our new story of a flourishing future? The natural sciences tell us of the profound empathic interconnectivity established through our common neuronal structure. Jungian psychology presents psyche as a dynamic process uniting us through its ability to express archetypal forms. These perspectives show that at a deep psychological level we gaze at the world through similar windows. The concept of archetypes arose through the commonalities in how we see and experience the world. They act as a psychological bridge, unifying us in our common gaze upon the world. We know we are in the presence of an archetype when we experience a profound, affective, numinous feeling in the moment of perception. The sense of awe in nature, in listening to a great composition, or standing before a great work of art unites us in universal reverie. In this moment we touch a psychological substrate and nod in unison at the thing experienced. In creating our new story we are building on archetypal patterns that bind us together and link us back to the deepest nature of our being in the world.

What of the philosophical threads that compose our tapestry of transformation? We must recognize the connections between our being and the world as integrally coupled. Who I am and how I define myself is infused with the many worlds surrounding me: personal, social, and political. I have considered the concept of our phenomenal world and how we can become more deeply connected to it by consciously shifting our sense of time, spatial relations, and body. These practices put us in touch with our ability to value the world and open us to an emerging ethic that unveils our capacity of creative caring for one another. The philosophic thread in our tapestry is grounded in our being in the world. How connected are we to the places in which we dwell, the bodies that are us, and the connections we have to others? In exploring these questions we begin to understand how the way we live affects the world.

This perspective on who we are and how we exist in the world rests on the observation that we are not separate beings. There is no way to cut the invisible cords tying us to our surroundings, and this becomes most apparent when our surroundings change. In moments of change we may find ourselves happy, sad, friendly, or angry. Outer circumstances truly define our being in many ways, and to view ourselves as separate entities in which we push the world around at will is an illusion. We push the world, and it reacts, but equally the world pushes us, and we react. We are in continual play with our many worlds, not only our physical world "out there" but also our social and economic worlds. We are always in participatory engagement with the world. I believe the most significant key to our transformation of being is remembering to ask ourselves questions; the philosophic thread will bring illumination only through the simple asking of questions about our being in the world.

The interweaving of these various perspectives provides us with a new way to envision our future. The specific forms arising from this tapestry are found in symbols appearing as images and metaphors. The transformative power of these forms cannot be understated; they represent archetypal forces locked within all of us. We create an innovative narrative that touches each of us through these symbols of transformation. This creative and animate story becomes the new Living Myth connecting us to one another and to Earth.

We have arrived at a time for transformation, a time to turn toward a future of unparalleled potential. This turning is an unfolding of an archetypal process of transformation allowing physical, biological, sociological, psychological, and philosophical qualities to manifest our new story of being in the world. Each of these perspectives opens our eyes to seeing the nature of the world as one. Our new story is one of global ecological emergence, providing us with a vision of the world as a single interconnected web in which we are co-creative contributors. Our challenge now is to work together to create this new story. We are not alone in our work of weaving our new Living Myth of connectivity and cooperation.

Each of us has the capacity to transform our lives.

Epilogue

As a scientist, I look at the world as objectively as possible. I rely on observations and the laws of physics to understand how Earth's climate has evolved through time and how it will change in the future. Science provides the *ground for understanding* how we are changing our planet. Human reasoning is exceptional for building the foundations upon which we can plan for our future, and I have enjoyed a rich, rewarding career in the sciences while learning much from my colleagues around the world. Many scientists believe that to move beyond the realm of reason threatens the validity of their work. I believe we need to expand our ways of knowing to include more than the logic of science.

I feel we need to enter onto a *path of compassionate action* to avoid the worst consequences of human-induced climate change. We cannot rely on technology alone to get us out of this situation; technology is only one aspect of the solution. If we open our hearts to the world's suffering and feel our connection to the world, our actions will be true. Our path to a flourishing future will succeed through compassionate action rooted in care for others. I do not become less of a scientist by opening my heart to the world. I become more whole. I embody a source of wisdom resting in each of us.

The *fruition of walking the path* of compassion and wisdom is a better world for generations to come. When I look at my daughters I see their future; I see the rocky path we have paved for them, one rooted in uncontrolled consumption. I want to create a different path that leads to a world of flourishing fruition for all. I encourage you to look at the younger generation and commit to walking on a new path leading to a world of wonder and beauty. Look within and follow the path of your heart. See the basic goodness within yourself and others. This goodness is the ultimate ground from which we all begin. Touch it and stay true to the path of compassion, and you will create that world of wonder.

Further Reading

Chapter 1. A Journey from Climate Science
to Psychology

Archer, D., and Rahmstorf, S. *The Climate Crisis: An Introductory Guide to Climate Change*. Cambridge: Cambridge University Press, 2010.

Haule, John Ryan. *Jung in the Twenty-First Century: Evolution and Archetype*. London: Routledge, 2011.

Henson, Robert. *The Thinking Person's Guide to Climate Change*. Boston: American Meteorological Society, 2014.

Jung, C. G. *Two Essays on Analytical Psychology*. Princeton, N.J.: Princeton University Press, 1977.

Kump, L., J. Kasting, and R. Crane. *The Earth System*. N.J.: Prentice Hall, 2009.

Mathez, Edmund A. *Climate Change: The Science of Global Warming and Our Energy Future*. New York: Columbia University Press, 2009.

Stein, Murray. *Jung's Map of the Soul*. Chicago: Open Court, 1998.

Stevens, A. J. *Jung: A Very Short Introduction*. Oxford: Oxford University Press, 2001.

Chapter 2. Learning to Embrace Change

Carney, D. R., et al. "The Secret Lives of Liberals and Conservatives: Personality Profiles, Interactions Styles, and the Things They Leave Behind." *Political Psychology* 29 (2008): 807–840.

Berry, Thomas. *The Great Work: Our Way Into the Future*. New York: Bell Tower, 1999.

Hillman, James. ed. *Puer Papers*. Dallas: Spring, 1991.

Jost, J. T., et al. "Political Conservatism as Motivated Social Cognition." *Psychological Bulletin* 129 (2003): 339–375.

Kahan, D. "Why Are We Poles Apart on Climate Change." *Nature* 488 (2012): 255.

Karen, Robert. *Becoming Attached: First Relationships and How They Shape Our Capacity to Love*. Oxford: Oxford University Press, 1998.

Laing, R. D. *The Divided Self*, London: Penguin, 1990.

Thomas, L., et al. *A General Theory of Love*. New York: Vintage, 2000.

Chapter 3. Facing Our Fears Associated with Climate Change

Bowins, B. "Psychological Defense Mechanisms: A New Perspective." *American Journal of Psychoanalysis* 64 (2004): 1–26.

Cramer, P. "Defense Mechanisms in Psychology Today." *American Psychologist* 55 (2000): 637–646.

Jacobi, J. *Complex, Archetype, Symbol in the Psychology of C. G. Jung*. Princeton, N.J.: Princeton University Press, 1959.

Jung, C. G. "A Review of the Complex Theory." In *The Structure and Dynamics of the Psyche*. Princeton, N.J.: Princeton University Press, 1981.

Kiehl, J. T. "A Jungian Perspective on Global Warming." *Ecopsychology* 4 (2012): 187–192.

Nesse, R. M., et al. "The Evolution of Psychodynamic Mechanisms." In *The Adapted Mind*, ed. J. Barkow et al. Oxford: Oxford University Press, 1992.

Randall, R. "Loss and Climate Change: The Cost of Parallel Narrative." *Ecopsychology* 1 (2009): 118–129.

Singer, T., and S. Kimbles, eds. *The Cultural Complex: Contemporary Jungian Perspectives on Psyche and Society*. London: Routledge, 2004.

Whitbourne, S. K. "The Essential Guide to Defense Mechanisms." *Psychology Today*. http://www.psychologytoday.com/node/77412.

Chapter 4. How Images Facilitate Transformation

Campbell, J. *The Inner Reaches of Outer Space: Metaphors as Myth and as Religion*. Novato, Calif.: New World Library, 2002.

Curry, P. *Defending Middle-Earth, Tolkien: Myth and Modernity.* Boston: Houghton-Mifflin 2004.

Hollis, J. *Mythologems: Incarnations of the Invisible World.* Toronto: Inner City.

Jung, C. G. *The Archetypes and the Collective Unconscious.* Princeton, N.J.: Princeton University Press, 1980.

Gottschall, J. *The Storytelling Animal, How Stories Make Us Human.* New York: Mariner, 2012.

Mandelbaum, A., trans. *The Metamorphoses of Ovid. Book VIII.* New York: Harcourt Brace, 1993.

Oreskes, N., and E. M. Conway. *The Collapse of Western Civilization: A View from the Future.* New York: Columbia University Press, 2014.

Pagel, M. *Wired for Culture: Origins of the Human Social Mind.* New York: Norton, 2012.

Chapter 5. Opposites and Our Relationship to Climate Change

Jung, C. G. *Psychological Types.* Princeton, N.J.: Princeton University Press, 1990.

Merchant, C. *The Death of Nature: Women, Ecology and the Scientific Revolution.* San Francisco: Harper, 1983.

Watts, A. W. *The Two Hands of God: The Myths of Polarity.* New York: Braziller, 1963.

Chapter 6. Balancing the Opposites of Climate Change

Dietz, T., et al. "Effects of Population and Affluence on CO_2 Emissions." *Proceedings of the National Academy of Sciences* 94 (1997): 175–179.

Gore, A. *Earth in the Balance: Ecology and the Human Spirit.* Boston: Houghton-Mifflin, 1992.

Harding, S. *Animate Earth: Science, Intuition, and Gaia.* Devon: Green Books, 2006.

Meadows, D. H. *Thinking in Systems: A Primer.* White River Junction, Vt.: Chelsea Green, 2008.

Oreskes, N., and E. M. Conway. *Merchants of Doubt.* New York: Bloomsbury, 2010.

Chapter 7. Exploring Our Being in the World

Dreyfus, H., and S. Kelly. *All Things Shining: Reading the Western Classics to Find Meaning in a Secular Age.* New York: Free Press, 2011.

Spinelli, E. *The Interpreted World: An Introduction to Phenomenological Psychology.* London: Sage, 1989.

van Manen, M. *Researching Lived Experience.* New York: SUNY Stony Brook Press, 1990.

Wrathall, M. A. *Heidegger and Unconcealment: Truth, Language, and History.* Cambridge: Cambridge University Press, 2011.

Chapter 8. Beauty's Way in the World

Abram, D. *The Spell of the Sensuous.* New York: Vintage, 1997.

André, C. *Looking at Mindfulness.* New York: Blue Rider, 2014.

Hayward, J. W. *Letters to Vanessa: On Love, Science, and Awareness in an Enchanted World.* Boston: Shambhala, 1997.

Jackson, T. *Prosperity Without Growth: Economics for a Finite Planet.* London: Earthscan, 2009.

O'Donohue, J. *Beauty: The Invisible Embrace.* New York: Harper, 2005.

Chapter 9. Why Meaning Is Important to Being in the World

Ford, D. *The Search for Meaning.* Berkeley: University of California Press, 2007.

Hollis, J. *Finding Meaning in the Second Half of Life.* New York: Gotham, 2005.

Jaffe, A. *The Myth of Meaning.* Einsiedeln: Daimon Verlag, 1984.

Kasser, T. *The High Price of Materialism.* Cambridge, Mass.: MIT Press, 2003.

Chapter 10. How Our Many Worlds Are Entwined

Capra, F. *The Web of Life: A New Scientific Understanding of Living Systems.* New York: Anchor, 1997.

H. H. the Dalai Lama. *The Universe in a Single Atom: How Science and Spirituality Can Serve Our World.* London: Abacus, 2007.

Chapter 11. Recognizing the Importance of the Transpersonal

Grim, J., and M. E. Tucker. *Ecology and Religion.* Washington, D.C.: Island, 2014.

Kaza, S., and K. Kraft. *Dharma Rain: Sources of Buddhist Environmentalism.* Boston: Shambhala, 2000.

Stanley, J., and D. R. Loy. *A Buddhist Response to the Climate Emergency.* Somerville, Mass.: Wisdom Publications, 2009.

Vaughan-Lee, L., ed. *Spiritual Ecology: The Cry of the Earth.* Point Reyes, Calif.: Golden Sufi Center, 2013.

Chapter 12. Awakening to One World

Capra, F., and P. L. Luisi. *The Systems View of Life: A Unifying Vision.* Cambridge: Cambridge University Press, 2014.

Karmapa. *The Heart Is Noble: Changing the World from the Inside Out.* Boston: Shambhala, 2013.

Sabini, M. *The Earth Has a Soul: The Nature Writings of C. G. Jung.* Berkeley: North Atlantic, 2002.

Index

abstraction, 61
advertising, 33, 109–10
ammonia, 62–63
anger, 29
animals, 47
Anthropocene epoch, 9, 24
anticipated loss, 34–35
anxiety: defense mechanisms
   for, 36–39; security and, 20;
   transformation and, 25; with
   anticipated loss, 34–35; with
   change, xi, 17–18, 20, 77–78;
   with climate change, 28–42;
   with death, 60; with world's
   end, 60
apartheid, 131–32
Apollo 8, 45
archetypes: complexes from,
   32; of disunion, 67–68; of
   feminine, 65–66; gods as, 108;
   of harmony seekers, 21–24;
   of hero and heroine, 50–54;
   interconnectedness of, 124;

on Internet, 90; in Jungian
   psychology, xii, 5, 19–21; of
   masculine, 64–66; meaning
   and, 108–9; of old person,
   19–21; of opposites, 59–60,
   65–68; of security, 21–24;
   transformation and, 136–37;
   of union, 67–68; within
   unconscious, 32; of young
   person, 19–21
Arctic sea ice, loss of, 28
Armstrong, Neil, 47
arts, 10, 53; intuition in, 62; meaning and, 110
attention deficit disorder, 89
Augustine (Saint), 135
authentic life, 98–99, 111–12
authority, 74–75; hierarchical, 21
autonomy, 35, 111–12
awakening: to interconnectedness,
   117–37; to One World, 138–42;
   transpersonal and, 129–37. *See
   also* transformation

151

Index

MAD. *See* mutually assured destruction
masculine, 65–66
Maslow, Abraham, 120–21
materialism: authentic life and, 111–12; autonomy and, 111–12; consumption and, 93; meaning from, 111, 113; myth of, 113
meaning: archetypes and, 108–9; arts and, 110; in beauty, 103; being in the world and, 105–14; complexes and, 107–8; consumption and, 111; creativity and, 106; emptiness and, 78; in feeling, 30; from gods, 112–13; from identification, 107; Jungian psychology on, 6, 107; lifestyle and, 109–10; from materialism, 111, 113; from power, 113; from religion, 106–7; security from, 106; from status, 109; symbols and, 47, 106; from transpersonal, xiii; from wealth, 113; in wholeness, 110–11
Merchant, Carolyn, 66
metaphor, 6, 53, 90; art and, 53; of Greeks, 49–50; for transformation, 55
methane, 63
Middle Ages, 75, 109
militarism, 107
mood, 95–96
morality, 6
mother, xii
mourning, 35
Muir, John, 121
multitasking, 89
music, 46, 124–26

mutually assured destruction (MAD), 130–31
myths: about creation, 60; about destruction, 60; of Greeks, 66, 112; Living Myth, 142; masculine/feminine dyad in, 66; of materialism, 113; about opposites, 60

nationalism, 107
natural resources, 78, 133
natural world, Greeks and, 48–50
Nature, as transpersonal, 135
nuclear war, 130–31

obesity, 93
oceans, 13, 92
O'Donohue, John, 99
Odyssey (Homer), 46
old person (*senex*), 19–21
one-sidedness, 58, 79
One World, 138–42
opposites: archetypes of, 59–60, 65–68; balance of, 72–82; of climate change, 57–82; contrast of, 60; creation/destruction dyad, 60; decision making and, 59, 64; Dis and, 69–70; emotions and, 70–71; of equality/authority dyad, 74–75; in future, 57; intuition and, 60–61, 62; masculine/feminine dyad, 65–66; myths about, 60; polarity of, 58, 71; psyche and, 75–76; in science, 58–59; senses and, 60, 61–62; transformation of, 71
overconsumption: balance and, 78–79; of energy, 78, 111; global